景观100+

LANDSCAPE DESIGN

商业 住宅 教育

Commerce Residence Education

ThinkArchit 工作室 主编

华中科技大学出版社
http://www.hustp.com
中国·武汉

前 言
Preface

本书收录了近年来全球优秀的景观设计案例，除了有大量案例的实景照片外，还配附了规划平面图、剖面图、手绘效果图以及翔实的文字说明等，深层次多角度地展现了设计者的设计理念和手法。编者按照景观功能将设计作品分为商业办公、住宅、教育、公园、广场街道、滨水六类。本书选取的项目既有简洁而精美的现代景观，又有富含深厚文化底蕴的古典式景观；既有恢宏壮观的东南亚式景观，又有那强调简洁、明晰的线条的欧式景观。

景观设计师用细腻的笔触勾勒出一幅幅与自然和谐共生的画面，它是激情和畅想，是隐逸和仁和，更是纯粹和完美。

This book includes the world's excellent landscape design cases of recent years. Apart from a large number of exquisite pictures, there are planning floor plans, section drawings, hand drawings and detailed descriptions in this book, presenting the concepts and design methods of landscape design in a profound way from multi–angles. The editor divides the design works into 6 chapters according to the landscape functions, i.e. Commercial and Official Space, Residence, Education, Park, Square and Street, Waterscape. There are not only concise and exquisite modern landscapes and classic landscapes with profound cultural connotations, but also magnificent south–eastern landscapes and european style landscapes set off by bridges, flowing water, pavilions, terraces and towers.

With fine and smooth brushwork, the landscape designers delineate pictures of human beings coexisting in harmony with nature. It is about passion and imagination, about seclusion, benevolence and harmony. It is more about purity and perfection.

商业办公 COMMERCIAL AND OFFICIAL SPACE

目录
Contents

住宅 RESIDENCE

教育 EDUCATION

100+ LANDSCAPE DESIGN

商业 办公

COMMERCIAL AND OFFICIAL SPACE

流浪者酒店

Vagabond Motel

设计公司：Lewis Aqüi Landscape + Architectural Design

地点： USA（美国）

面积：0.53 hm^2

摄影：Lewis Aqüi, Orlando Rio

材料：South Florida Native Planting Palette,
Concrete "Sea Shell" pavers for walkways and
pool deck and pool coping, IPE wood decks

Lewis Aqüi Landscape + Architectural Design (LA^2 d) is the designer of the exterior renovation of the iconic Vagabond Motel.

This classic motel depicts all the characteristics of the Miami Modern style (MIMO Architecture), including an open-air plan, geometric designs, angular planes, overhanging roof lines and "Eyebrows", natural stone veneering, and sculptural elements and focal points depicting marine life and other nautical themes typical of the era. The Motel was originally designed in the 1950's by B. Robert Swartburg who was one of the most prominent and innovative architects in Miami-Dade County. He created other landmark structures as the Miami Civic Center Complex and the Delano Hotel.

Lewis Aqüi has recreated the outdoor spaces using the planting palette that is primarily native to South Florida and which was popularly used in the mid-century designs. At the same time Lewis kept in mind a planting design that is "Florida Friendly" which conserves water, protects the environment, adaptable to local conditions and is drought tolerant.

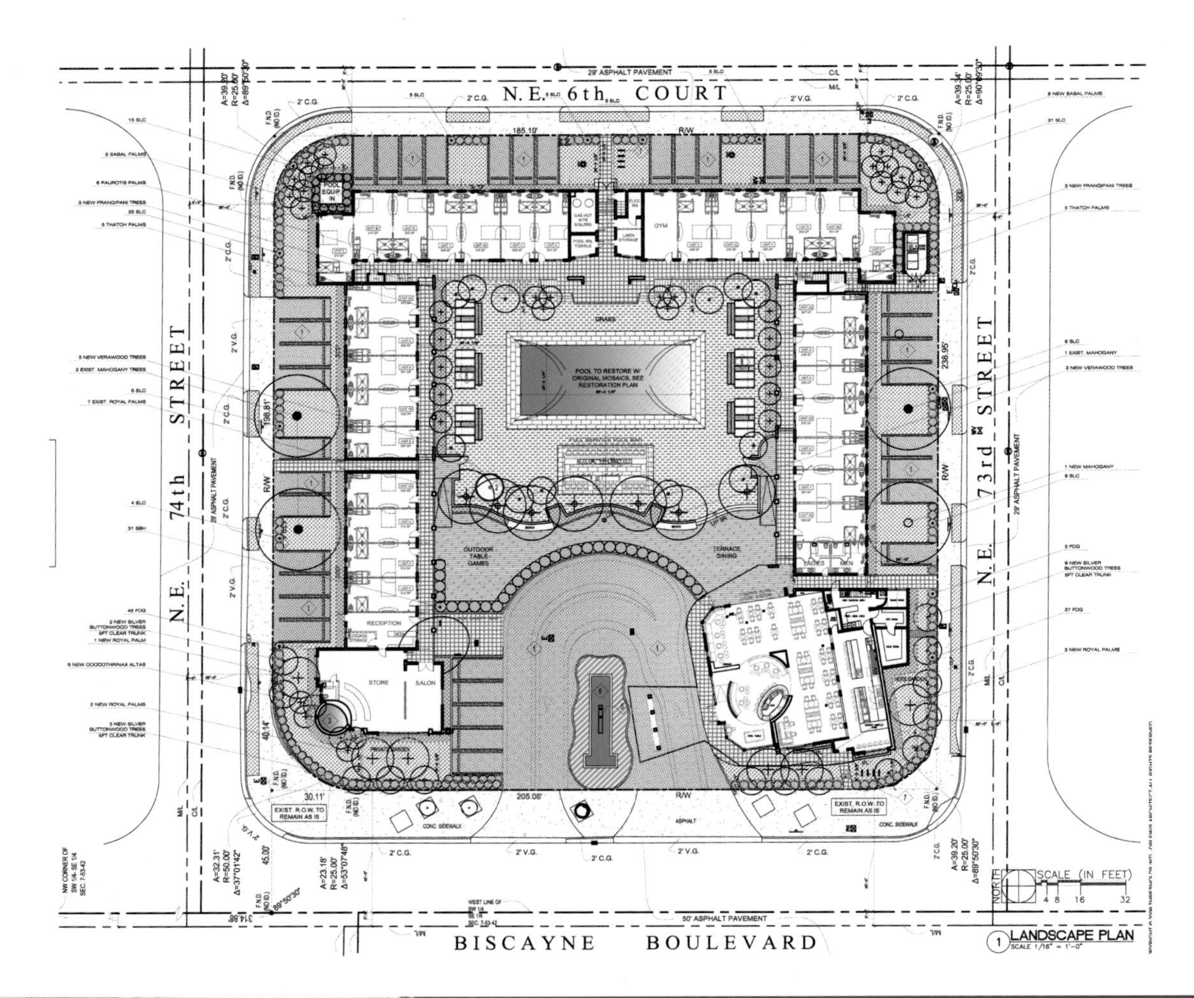
N.E. 6th COURT
N.E. 74th STREET
N.E. 73rd STREET
BISCAYNE BOULEVARD
POOL TO RESTORE W/ ORIGINAL MOSAICS, SEE RESTORATION PLAN
OUTDOOR TABLE GAMES
TERRACE DINING
RECEPTION
STORE
SALON
GYM
SCALE (IN FEET)
LANDSCAPE PLAN

VAGABOND MOTEL

VAGABOND MOTEL

本项目由 Lewis Aqüi Landscape + Architectural Design (LA^2 d) 建筑设计工作室设计完成。

这个经典的汽车旅馆采用了迈阿密现代风格，包括户外设计、几何设计、平面设计、屋顶线设计、石材镶边设计、结构元素设计等，描绘了海军生活和航海时代的生活。汽车旅馆原先于 20 世纪 50 年代由 B. Robert Swartburg 设计完成，他是迈阿密戴德郡突出的优秀的设计师。他设计了迈阿密市政中心大楼和德拉诺酒店等标志性建筑。

Lewis Aqüi 建筑设计工作室重新设计了户外空间，选择一些具有南佛罗里达州当地特色的植物做搭配，这些搭配常见于中世纪的装饰空间中。同时，建筑设计工作室采用“温暖的佛罗里达”的植物设计方案，起到节约水源、保护环境、耐旱的作用。

半岛度假酒店

Peninsula Resort

设计公司：Quid Studio

设计师：SArchitettura Sonora

地点： Italy（意大利）

面积：1440 m^2

摄影： Emilio Bianchi

After more than thirty years the disco club Seven Apples and the contiguous Peninsula Beach resort is confirmed as one of the legendary venues along the Versilia Riviera, in the Tuscan coast between the celebrated Forte dei Marmi and the beautiful Pietrasanta.

A fabulous and celebrated location, Seven Apples has always been able to constantly renew itself, remaining faithful to a precise idea of excellence, fun and class.

To pursue such an idea, the Peninsula Beach Resort, the outdoor area of the club, has provided itself with an incredible outdoor space powered by an Architettura Sonora sound system that makes people feel immersed into music and fun.

Such a sophisticated system, with the highly directional A.S. sound modules, has also been a winning choice in terms of sound dispersion, minimizing any kind of noise disturbance outside the club boundaries.

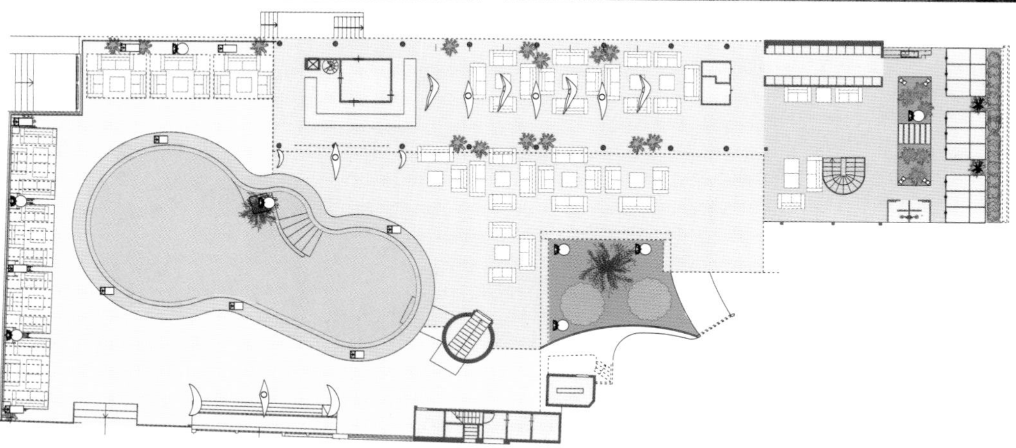

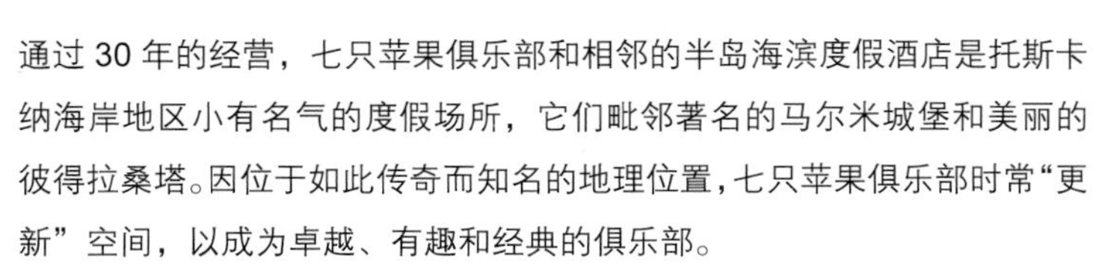

通过 30 年的经营，七只苹果俱乐部和相邻的半岛海滨度假酒店是托斯卡纳海岸地区小有名气的度假场所，它们毗邻著名的马尔米城堡和美丽的彼得拉桑塔。因位于如此传奇而知名的地理位置，七只苹果俱乐部时常“更新”空间，以成为卓越、有趣和经典的俱乐部。

为追求卓越、有趣、经典的主旨，设计师为俱乐部的门户——半岛海滩度假酒店设计了全新的形象，并为游客提供了难以置信的精彩体验，其中包括索诺拉建筑音响系统，它能令人们仿佛置身于音乐的海洋。

索诺拉建筑音响系统搭配高质量的 A.S. 声音模组创造出了无与伦比的音响效果，并能够减弱俱乐部边界以外任何类型的噪声干扰。

滨海度假村

A Coastal Retreat

设计公司：Francis Landscapes

地点：Lebanon（黎巴嫩）

面积：2 hm^2

Extending along a kilometer of the Lebanese southern coastline, this exclusive beach resort is nestled among exotic guava groves and is designed to open up on an untamed but tranquil sea and a magnificent setting sun. Gently slanting towards the sea, winding pathways invite visitors to a contemplative walk or a refreshing swim in the sea. The organic aspects of this project grew from the dedication to make it as eco-friendly as possible, carving the entire landscape around the already existing vegetation, fully aware that luxury lies within nature.

With its paradisiacal beach, the lush greenery and exotic fruiting trees surrounding this stretch of beachfront property made this project exotic from the onset. To create a true natural haven and complete the existing Eden-esque atmosphere, raw materials, such as Ylang Ylang, were shipped in from Bali.

A strong, early coordination between the architectural and landscaping work allowed for an ideal outcome, with nature at its core. The soft landscape made up of lawns, exotic and flowering trees, and shrubs, all seem to engulf the hard landscape made up of pools, kids pools, sundecks, bungalows, and restaurants scattered across the resort and positioned to give beach goers an idyllic view from every angle.

The true jewels of this project however, are the personal, serviced bungalows, with private jacuzzis overlooking the sea. Tailored garden inspires comfort and escape. From these bungalows, we created a space that whisks visitors away to a destination far from the hustle and bustle of Beirut.

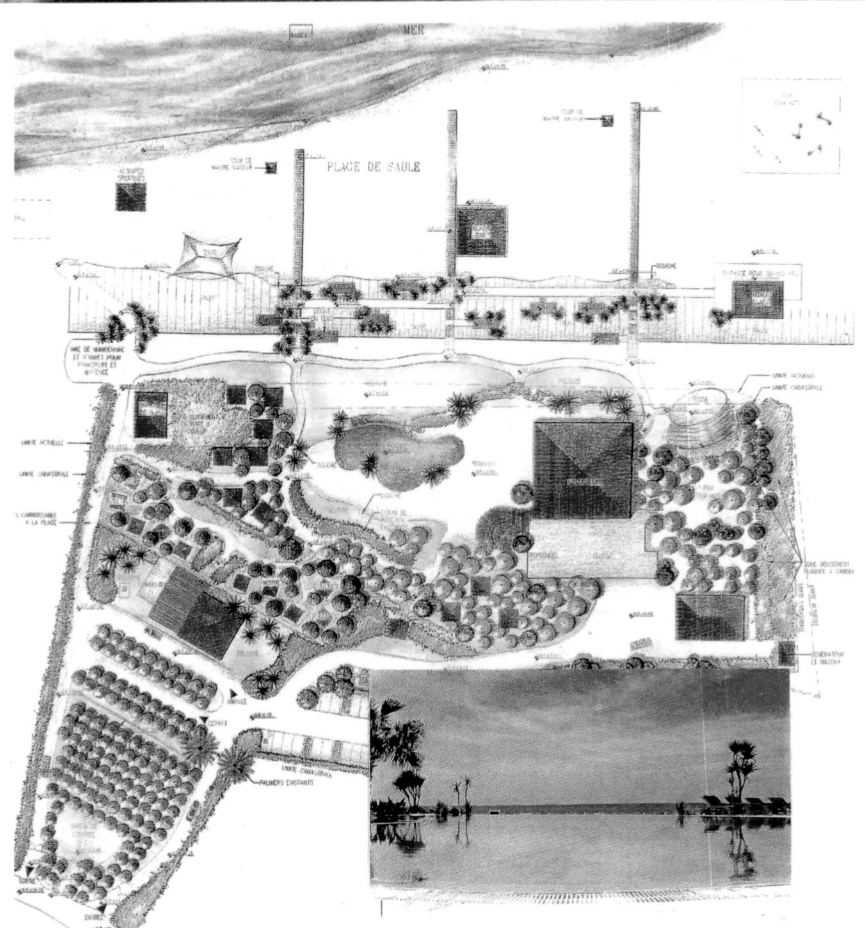

MER
PLAGE DE SABLE

黎巴嫩南部海岸 1km 内有一处开放的海滨度假胜地，周围环绕着充满异国情调的番石榴树林。这里的海面平静、夕阳壮丽，是一片未经人工开发的自然浴场。游客们沿着蜿蜒的小径来到海边，或是漫步、或是畅爽地游泳。项目十分亲近自然，整个景观位于原始丛林中，充分体现了奢华与自然的统一。

项目拥有沙滩和美丽的海景。郁郁葱葱的树林围绕着海滨酒店。为了营造一个真实的、自然的伊甸园般的氛围，设计师从巴厘岛运来了香油树等作建筑材料。

项目以自然为核心，结合农业和景观，创造了一个理想之境。软景观包括草地、草花、树木、灌木，它们种植在硬质景观的周围，包括水池、小孩水池、太阳板、别墅、景区餐厅——给沙滩上的游客带来了田园风光。

项目的奢华之处在于有可提供个人服务的别墅，私人的洗浴中心可以鸟瞰大海，私人定制的花园能打造出舒适的休憩空间。这些别墅是游人们远离贝鲁特城市喧嚣的理想的目的地。

GRIZZA

贝尔蒙德帕拉西奥拿撒勒库斯科宫酒店

Belmond Palacio Nazarenas Hotel

设计公司：Rafael Rivero Terry Landscape Architecture

地点：Peru（秘鲁）

面积：5418 m^2

摄影：Rafael Rivero Terry

Belmond Palacio Nazarenas Hotel is a boutique hotel which was built in an old colonial monastery located in the Inca city of Cusco, Peru. The monastery was built on an Inca Temple.

It has tried to rescue the functional essence of what were the cloisters in the colonial era.

The hotel has several indoor gardens:first cloister, the second cloister or secret garden (garden of herbs), the garden of native plants, the garden of fruit trees or the orchard garden, the water garden,the pool garden, hallways.

Two of the main features of these gardens are its utilitarian and ornamental uses. Utilitarian for having many orchards and herb gardens that are used by the hotel chef and, as well as the variety of flowers marking an explosion of color and aromas.

贝尔蒙德帕拉西奥拿撒勒库斯科宫酒店是一个精品级酒店，位于曾经的殖民地——秘鲁的印加城库斯科，其原址是一座古庙。

酒店设计旨在复原殖民地时期古庙功能的精髓部分。

酒店里有一些内院：前寺院亭、中寺院亭（也叫秘密花园、香草花园），以及本土植物园、果园、带池子的花园、过道。

这些花园具有的两个主要特色是实用性和观赏性。香草和果子可以被酒店大厨做成食物，各种香味和颜色的花可用来观赏。

Ruda
Rue
Ruta graveolens

皮克林花园酒店

Park Royal on Pickering

设计公司：TIERRA DESIGN (S)

设计师：Mr. Franklin Po

地点：Singapore（新加坡）

面积：6960 m^2

摄影：Amir Sultan

Located within the Central Business District and Chinatown areas, adjacent to Singapore's first privately owned garden – Hong Lim Park, Park Royal on Pickering is an example of how 200% Green (as defined by Building and Construction Authority, Singapore) can be achieved in a city that is very conscious of its space usage. The primary concept was to demonstrate how greenery can be conserved in a way that integrates harmoniously with the form and function of a business hotel and office development while seamlessly combining different aspects of design including architecture, landscape and interior design.

The visitor's experience begins at the street level, where tropical greenery complemented by architectural elements blur the lines between the public realm and the hotel. Reflecting pools and sculptural flora at the entrance make for a gentle and cool transition from the warm urban environment. The integration of the sidewalk and the building setback into the design, bringing greenery into the hotel lobby and the restaurant, creates a spatial quality of being in a garden.

A visually striking contoured podium is sculpted with softscape to create outdoor plazas, walkways and gardens which flow seamlessly into dramatic interior spaces. The cascading verdant contours conceal the raised car parking, melding into lush openings, crevasses, gullies and waterfalls thus creating an attractive urban element. The lofty sky gardens bring greenery to the rooms and internal spaces; corridors, lobbies and common areas are treated with

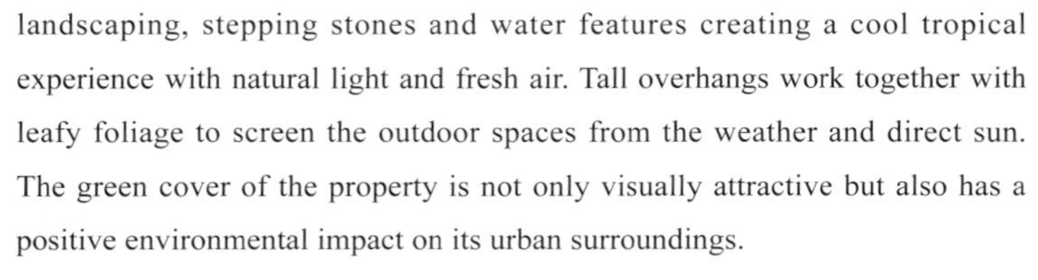

landscaping, stepping stones and water features creating a cool tropical experience with natural light and fresh air. Tall overhangs work together with leafy foliage to screen the outdoor spaces from the weather and direct sun. The green cover of the property is not only visually attractive but also has a positive environmental impact on its urban surroundings.

The general palette of plants and trees are chosen to create a tropical garden with hints of the exotic. In addition, the planting strategy focuses on engaging with the surroundings as much as possible. The greenery of the Hong Lim Park is drawn up the building in the lush planting of the podium and sky gardens. These four storey high sky gardens are the city's first zero energy sky terraces, carefully planted with low maintenace flora that are not only aesthetically important but also functionally critical in mitigating thermal gain on the west-facing wall and improving indoor air quality. In a city that is very conscious of its water usage and resources, the external areas of the hotel including the hardscape and the softscape become surfaces for rainwater collection. Calculations based on average rainfall indicate sufficient collection for irrigation of the entire podium area.

花园酒店位于人口密集的中央商务区和唐人街，毗邻新加坡第一个私有花园——芳林公园，实现了 200%的绿化率（由新加坡建设局测定），在这个寸土寸金的城市堪称典范。该项目完美整合了建筑、景观、室内装饰等不同领域的设计，实现了商业酒店、写字楼的形式与功能和谐统一，实现了新加坡建设“花园城市”的愿望。

访客体验从街道开始，热带植物辅以建筑元素使酒店融入公共空间。入口处的水池和造型植物营造出一片温柔和凉爽，让人顿时摆脱了城市的燥热。人行道将绿意一路带进酒店大堂和餐厅，令人仿佛身处花园。

抢眼的波浪形裙楼将户外广场、人行道和花园与内部空间无缝衔接。层叠状的轮廓造型遮蔽了停车场，密布的裂缝、沟壑和瀑布独具魅力。空中花园为内部空间带来绿色。走廊、大堂和公共区域步步皆景。石阶、水景、阳光、空气为热带风光增添了趣味。高大茂密的植被遮挡了直射的阳光，美观又环保。

这座热带花园充满异国情调，尽可能与周边环境融为一体。4 层楼高的空中花园是全市首个零能耗空中露台，精心种植着维护简单的花草，不但景致优美，也减少了西晒墙面的热量，改善了室内空气质量。新加坡非常注重节约用水，花园酒店外部区域所有地表均为雨水收集面。以平均降雨量计算，其收集量足够为整个裙楼的植物提供灌溉用水。

西班牙酒店

Castell D’emporda

设计公司：Concrete

设计团队：Erikjan Vermeulen, Rob Wagemans, Cindy Wouters, Melanie Knuewer

地点：Spain（西班牙）

面积：250 m^2

摄影：Ewout Huibers, Robert Aarts

Hotel Castell D’emporda located in Girona, Spain offers a signature restaurant including a large terrace with great views over the surrounding landscape. Concrete designed, at the clients’ request, a roof or covering for this terrace with the possibility to create an enclosed space with full wind and rain protection. One of the design conditions was to create a covering that works in harmony with the historical and listed building. Additionally we wanted to maintain the terrace feeling while be seated under the covering.

In principle a terrace is an outdoor space where one can enjoy the weather. If necessary, you need a parasol for sun or rain protection, but there is almost no obstruction between the visitor and the view. The solution was to create abstract parasols. 12 circles in divers diameters are placed randomly on the terrace. Where the circles touch they melt together, the open spaces between circles are filled in with glass. The circular parasol shapes enhance the feeling of being in an outdoor environment on a terrace. A glass roof or a winter garden would to much become a building, create a feeling being inside a structure and would also appear as an extension of the building, damaging the ancient character.

The top and edge of the parasols are made in rusted steel, seeking harmony with the ancient building and the natural environment. The white painted steel columns and ceiling create an open and light outdoor atmosphere under the parasols. Transparent sliding curtains can be hung easily in colder periods but always stay open. When the mistral winds suddenly appear the whole terrace

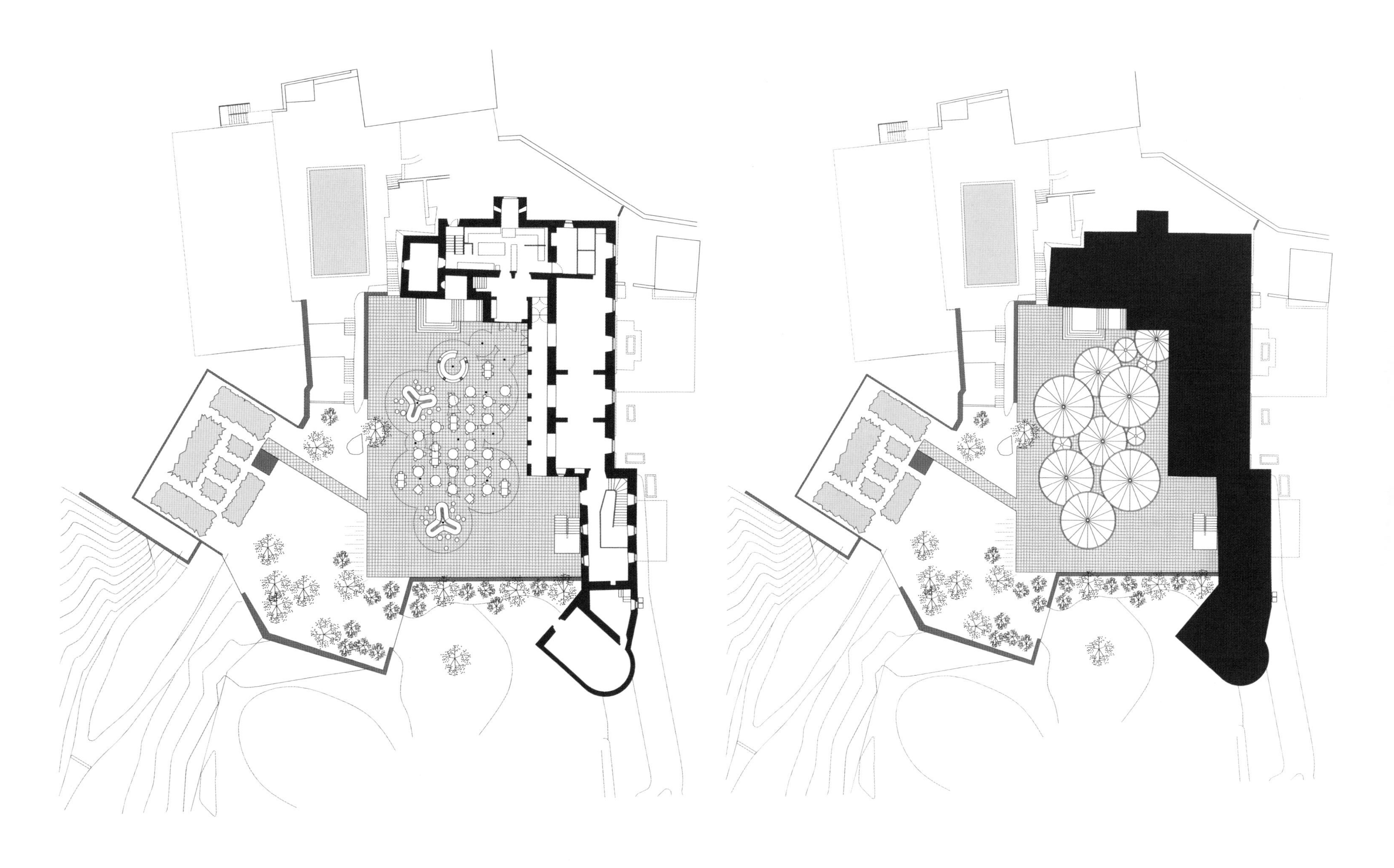

can be closed in a couple of minutes. Round and square marble tables and two white leather lounge couches create different seating facilities.

西班牙酒店外形十分显眼，在这里可以饱览周围美丽的景色。客户希望建筑师为他们设计一个混凝土材质的、相对封闭的、遮风挡雨的屋顶或挡板，“屋顶”要与历史建筑和谐统一，并且保留了露台空间。

一般来说，露台通常是一个供人们享受自然的户外空间。必要时，人们需要一个遮挡风雨和烈日的大伞，而且伞不能影响人们欣赏风景。建筑师想到的解决办法是，建造抽象的伞的形象。建筑师将 12 个“圆形伞”随机放置在露台上，圆形相互交叠的部分完美融合，不相交叠的部分则补上玻璃材料。如果将构筑物全部处理成玻璃屋顶，就会像温室那样过于封闭，从而破坏了构筑物与古建筑的融合感。

构筑物的顶部与侧边做了生锈的效果，与古建筑完美搭配，下部则采用了白漆刷成的钢柱，营造出开放明亮的户外空间。构筑物可以保持开敞状态，在寒冷的天气或起风时，人们也能快速地拉上透明的窗帘。构筑物里面摆放着精心设计的圆形和方形的大理石桌，还有白色皮革沙发供人们休息。

2015 米兰世界博览会巴西馆

Brazil Pavilion - EXPO Milano 2015

设计师：Studio Arthur Casas and Atelier Marko Brajovic

地点：Italy（意大利）

面积：plot area 4 133m^2, built area 3 674m^2

摄影：Raphael Azevedo França, Atelier Marko Brajovic

Studio Arthur Casas and Atelier Marko Brajovic won the competition to create the Brazilian Pavilion for Expo Milan 2015. We aimed to combine architecture and scenography in order to provide visitors with an experience that would transmit Brazilian values and the aspirations of its agriculture and livestock farming according to the theme "Feeding the world with solutions". More than a temporary building, the sensorial immersion includes leisure, high technology information, interaction and learning.

The inspiring idea of a flexible, smooth and decentralized network is present in every aspect of the building and represents the country's pluralism. Amidst more than 130 constructions, the Brazilian Pavilion proposes a pause, the intention of creating a public square that draws people together and engenders curiosity. As porous as the Brazilian culture, a large volume is open to visitors and establishes a pathway among several plant species cultivated in our country. The earthly colors of the metal structure highlight this "Brazilianess" and the gradual transition between inside and outside erase the boundaries dividing architecture and scenography. The metaphor of the net is materialized by a tensile structure that creates unexpected places for leisure and rest. Following the tradition of Brazilian modernism and its pavilions.

Different themes inspired the clusters distributed along the ground floor of the pavilion. They are organized according to ideas such as nutrition,

family agriculture, forestry and integration between farming and livestock. Sustainability is everywhere, from the construction/deconstruction system made up with prefabricated modules, to the water reuse mechanisms and the employment of certified and recyclable materials. The rationality of this ephemeral architecture demonstrates that it is possible to create meaning and content with few resources and low environmental impact.

The Brazilian pavilion in Expo Milan 2015 aims to bring new elements to the traditional attendance of the country to this type of event. Looking at the future, it aims to demonstrate that Brazil achieved excellence in crucial areas for mankind, such as agriculture and livestock farming, in a permanent movement to create new paradigms for the way our society relates to the environment, a symbiotic transformation, capable of tracing new strategies for our country. It should demonstrate that it is possible to transform into reality utopian ideas and to inspire solutions that follow the Expo theme: Feeding the planet, energy for life.

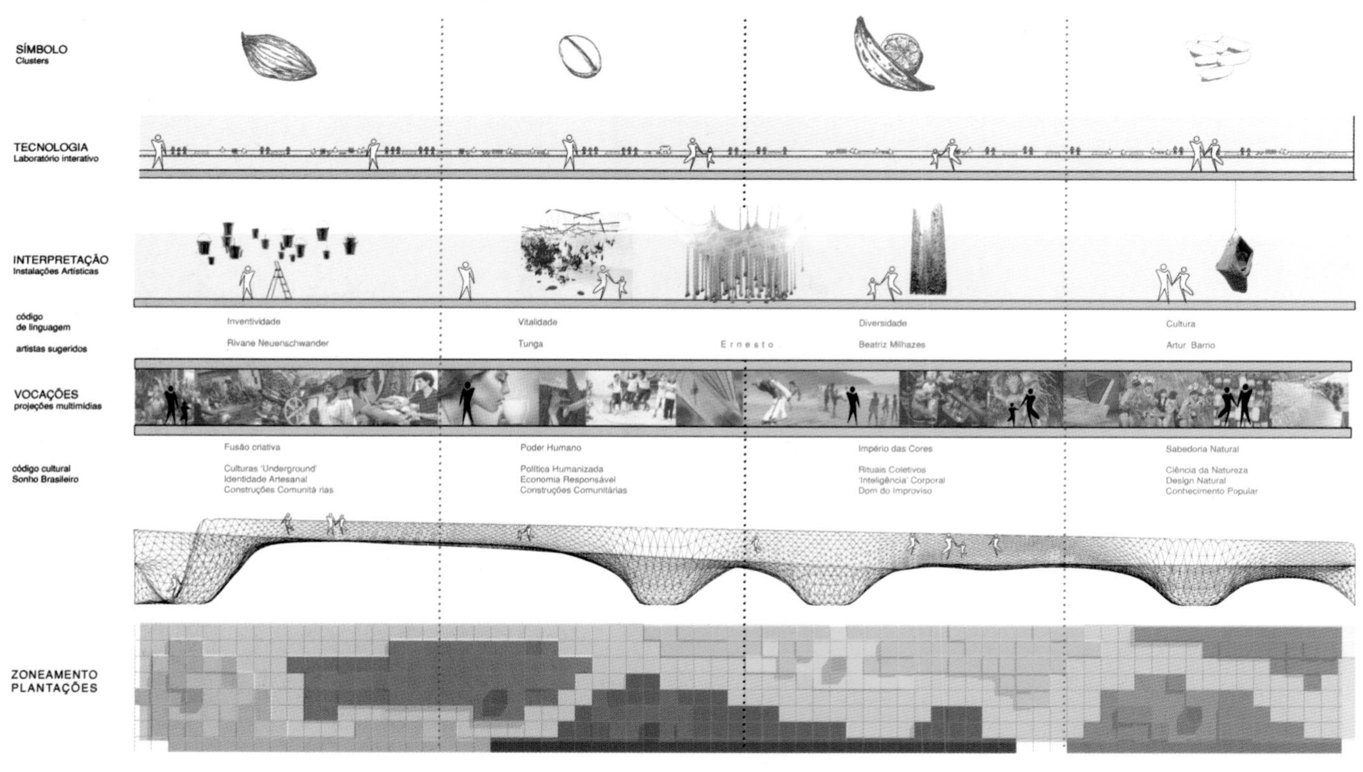
SÍMBOLO
Clusters
TECNOLOGIA
Laboratório interativo
INTERPRETAÇÃO
Instalações Artísticas
código
de linguagem
artistas sugeridos
Inventividade
Rivane Neuenschwander
Vitalidade
Tunga
Ernesto
Diversidade
Beatriz Milhazes
Cultura
Artur Barrio
VOCAÇÕES
projeções multimídias
Fusão criativa
Culturas 'Underground'
Identidade Artesanal
Construções Comunitá rias
Poder Humano
Política Humanizada
Economia Responsável
Construções Comunitárias
Império das Cores
Rituais Coletivos
'Inteligência' Corporal
Dom do Improviso
Sabedoria Natural
Ciência da Natureza
Design Natural
Conhecimento Popular
código cultural
Sonho Brasileiro
ZONEAMENTO
PLANTAÇÕES

Arthur Casas 和 Atelier Marko Brajovic 建筑工作室赢得了设计 2015 年米兰世界博览会巴西馆的机会。设计师计划将建筑与透视结合。由于本届展会的主题是“滋养地球，为生命加油”，因此，建筑的设计理念也源自于巴西的农业和畜牧业。建筑不是临时搭建的，它具有娱乐、高科技、互动和学习的功能。

建筑具有灵活、流畅、分散的特点，体现巴西农业的多样化。在 130 多个场馆中，巴西馆是一个能让人们停下来花几分钟时间研究，并且好奇这是个什么东西的建筑。一个大型的入口向游人们开放，道路旁栽种着巴西本土的农作物。建筑架构使用大地色的金属材料，强调“巴西特色”，从内到外的渐变，消除了建筑与透视的界限。网格呈拉伸的形态结构，超越传统形式，构造出娱乐区和休息区。巴西馆符合巴西的现代建筑美学。

展馆的一层按照不同主题分布有不同的展区，主题包括营养区、家庭农业区、林业区、农业区和畜牧业区等，这里大大小小的种植槽里种植了多种多样的植物，形成丰富的植物景观。

可持续发展的理念贯穿在整个设计中，从模块的安装、拆除系统，到再生水系统和可循环材料的使用。这种临时的建筑证明用少量材料建造出有内涵、有意义，并且对环境影响很低的建筑是可行的。

2015 米兰世界博览会巴西馆旨在把新的元素融入传统的展会上。展望未来，它将证明巴西在人类社会的重要领域——农业和畜牧业上获得的斐然的成绩。巴西馆在环境、再生改造、新策略等方面为社会创造新典范。

2015 米兰世界博览会法国馆

French Pavilion - EXPO Milano 2015

设计公司：XTU ARCHITECTS

设计师：Anouk Legendre & Nicolas Desmazières, Mathias Lukacs

地点：Italy（意大利）

面积：3 500 m^2

摄影：Andrea Bosio, Luc Boegly

France symbolises a cultural wonder, industrial know-how, the good life, reflects architect Anouk Legendre. That is what we wanted to show the world by inventing a built landscape that all at once portrays the geographic diversity of France's regions, its unique agricultural offerings and culinary traditions. When asked about the theme "Feeding the planet, energy for life", XTU says, Terrain as fertile ground for the new food revolution, with a building that stands for the promise of France's regions. The pavilion, inspired by France's hexagonal shape, seems to have been pushed up here and there by tectonic shifts. This built landscape has sidled into the market on the underside of the ceiling, the only part crowds will see as they stream into the 2,000 m² space. The landscape ceiling casts a striking feature that abstractly depicts the wide breadth of France's territories. That is what introduces the scientific content staged by Adeline Rispal's exhibition design.

法国是一个象征着文化奇迹、发达工业技术和美好生活的国家。设计师通过创造一个建筑中的室内景观，向世界展示法国区域的多样性、独特的农产品及烹饪传统。当谈论起设计主题“滋养地球，为生命加油”时，设计师说，土地是食物革命的基础，法国馆代表了法国地区的希望。建筑的设计受到法国六边形形状的启发。这一建筑占地面积为 2 000m²，人们进入场馆后会被建筑顶棚的设计所吸引。这也是展览设计要展现的科学内容。

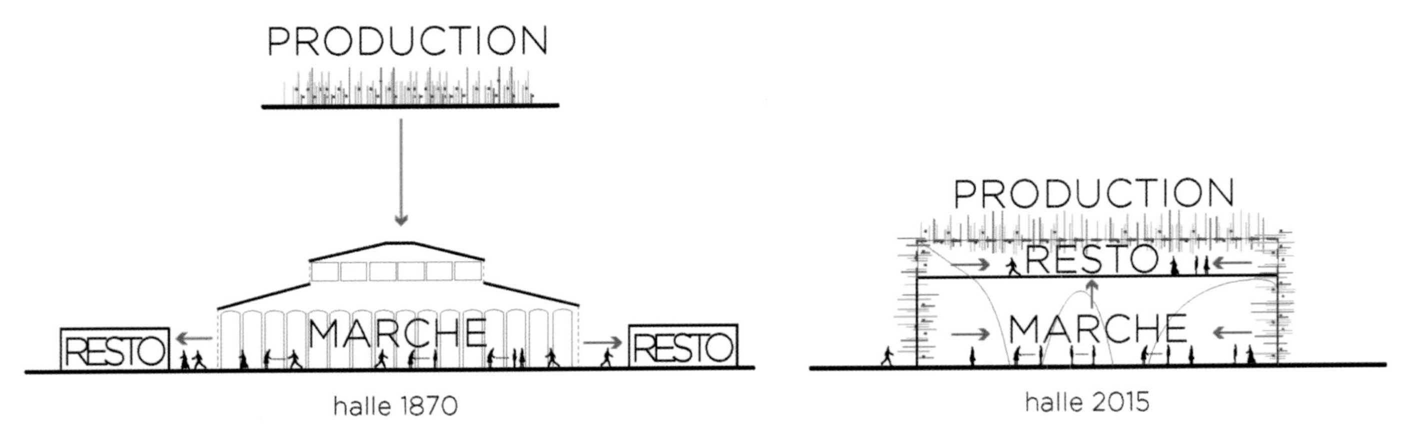
PRODUCTION
RESTO
MARCHE
RESTO
halle 1870
PRODUCTION
RESTO
MARCHE
halle 2015

PLAN DE MASSE ECHELLE : 1/500
PAVILLONS THEMATIQUES
PAVILLON ITALIA
PAVILLON
PAVILLON ISRAELIEN
PLACETTE
PAVILLON
PAVILLON VATICAN
DECUMANUS
ENTREE PUBLIC PAVILLON FRANCAIS

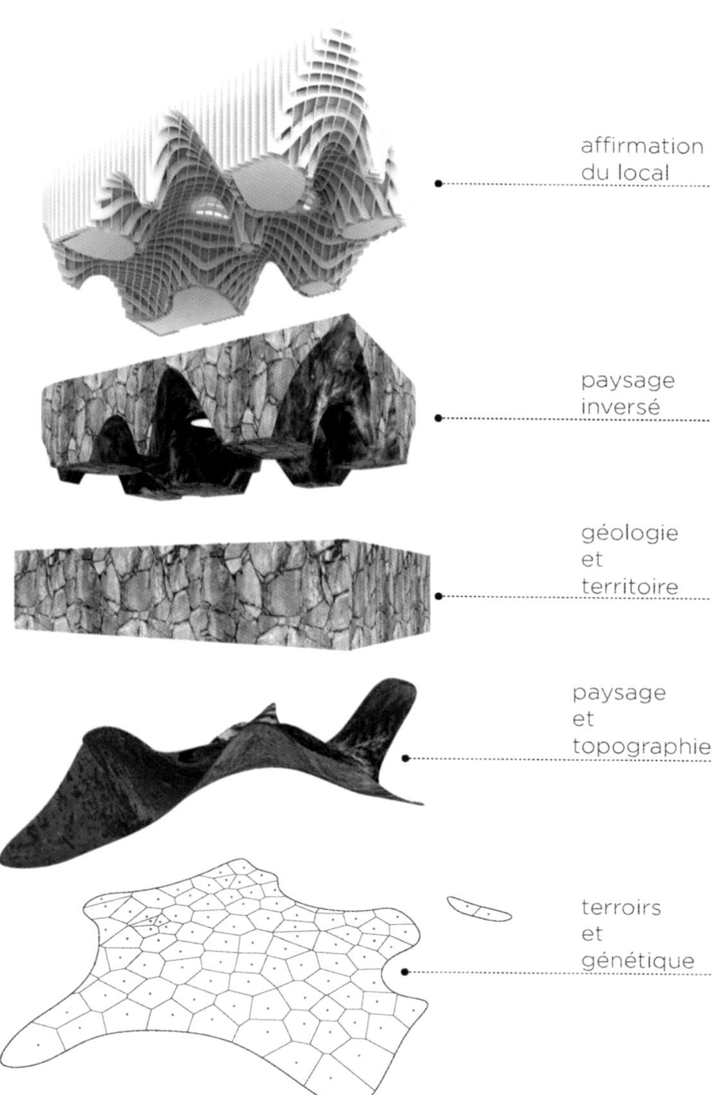

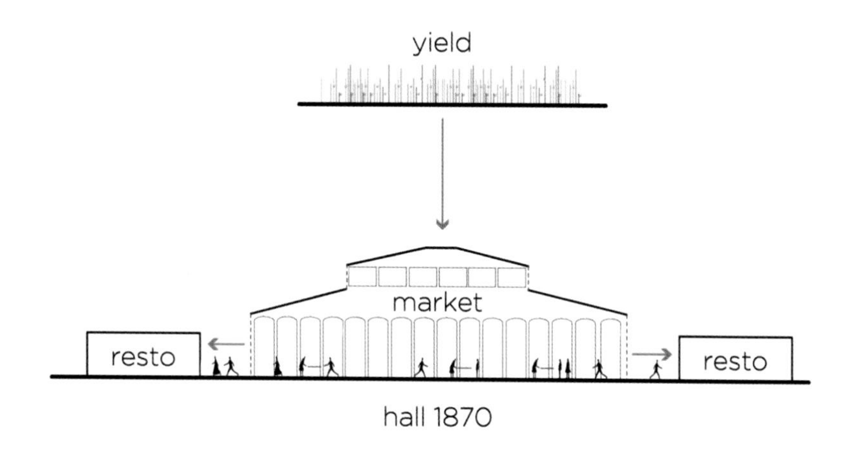

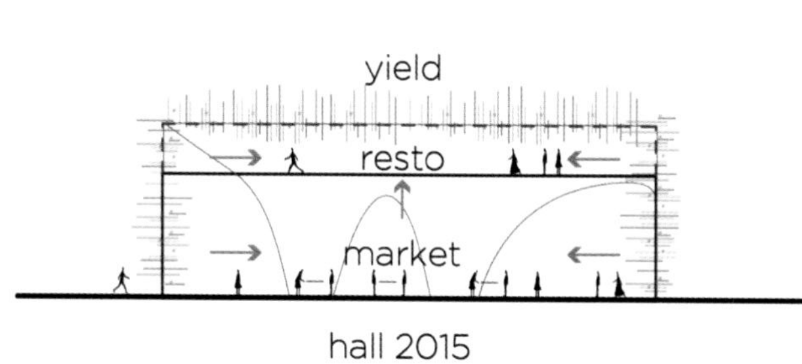

城市天空农场

Urban Skyfarm

设计公司：Aprilli Design Studio

设计师：Steve Lee

地点：South Korea（韩国）

面积：157 800 m^2

(Sky farm 144 450 m^2, Hydroponic Farm 13 350 m^2)

The Urban Skyfarm is a vertical farm design proposal for a site located in downtown Seoul, right adjacent to the Cheonggyecheon stream which is a heavily populated dense urban area within the central business district. Inspired by the ecological system of giant trees.

Through lifting the main food production field high up in the air, the vegetations gain more exposure toward the natural sunlight and fresh air while the ground level becomes more freed up with nicely shaded open spaces which could be enjoyed by the public. The bio mimicry of the tree form gives many structural and environmental advantages to form a light weight efficient space frame which could host diverse farming activities. The four major components which are the root, trunk, branch and leaf each have their own spatial characteristics which are suitable for various farming conditions.

The lower portions enclosed by the structural skin provide more controlled environments for solution based leafy productions. During daytime the photovoltaic panels generate electricity to be used for night time lighting and heating to support farming. The Urban Skyfarm creates a mini ecosystem which brings balance back to the urban community.

The Tree-like form creates a strong iconic figure in the prominent location and becomes a symbol of well being and sustainable development. Together with the Cheonggyecheon stream, the Urban Skyfarm will become a nice destination place for people seeking for fresh food, air and relaxation within their busy urban life.

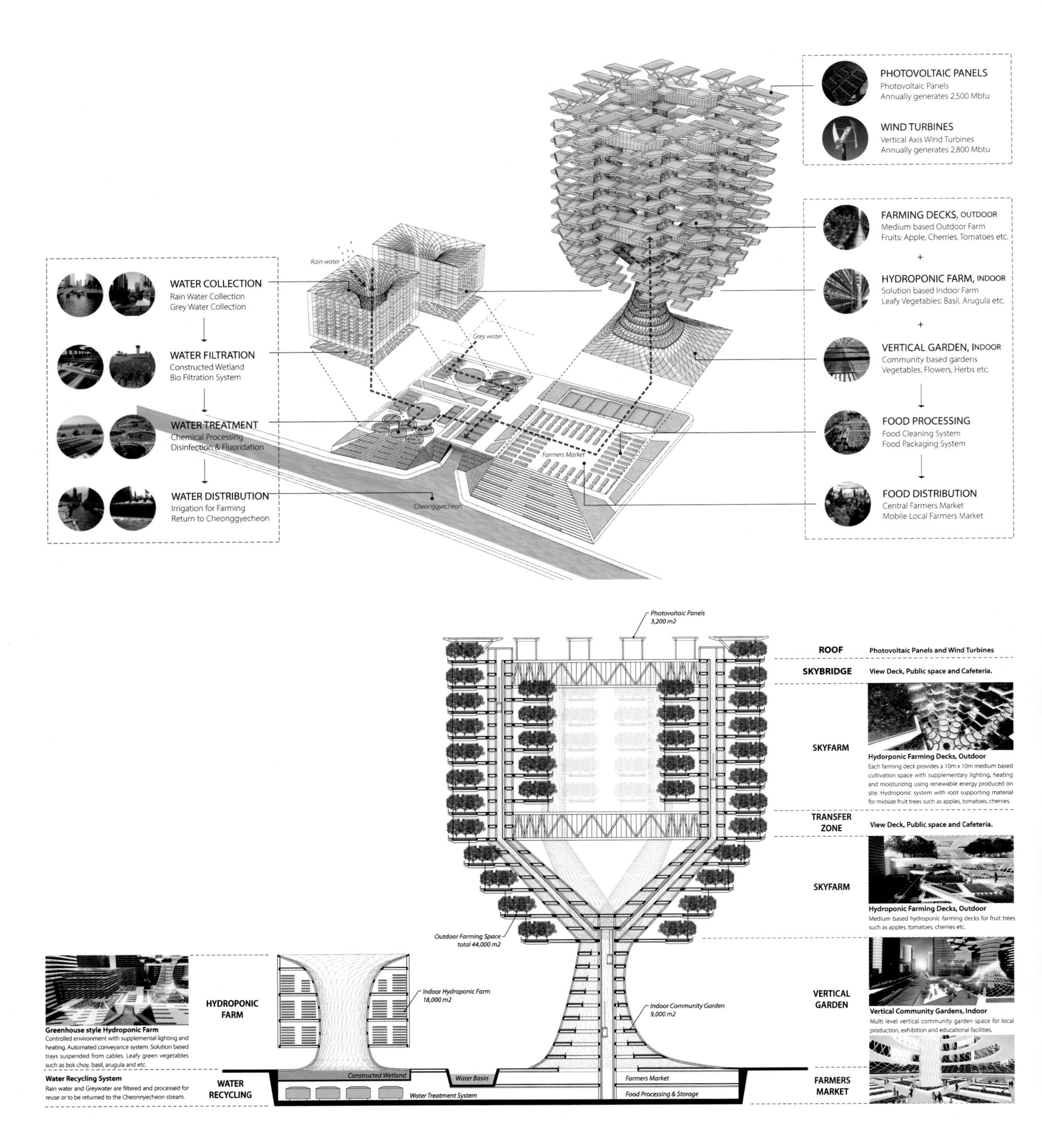

城市天空农场是一个垂直的农场设计方案，项目位于韩国首尔，毗邻清溪川，是韩国人口最密集的商业区之一。城市天空农场的设计灵感源自巨大树体的生态系统。

通过抬高农作物的用地，蔬菜得到了更多的自然阳光、新鲜空气，此外，地面土地就空置出来了，精心打造的高空空间是一种全新的建筑结构模

式。仿生学的树形结构在结构上和环境上创造了重量轻、效率高的空间架构，可以满足承载多样农作物的需求。建筑的四个主要结构是根部、树干、树枝、叶子，每个都有自己的空间特色，适用于多种农产品生长。建筑的低矮部分外围覆盖有一层建筑表皮，提供了可调节的环境，有助于农作物的生长。白天，太阳能光伏板聚集的热能在夜间可用于农场的照明和加热。城市天空农场创造了一个迷你的生态系统。

树状的外形及建筑的生态功能，使建筑成为特色、健康、可持续发展的代表。在清溪川的映衬下，城市天空农场成为人们在忙碌的都市生活中寻找新鲜食物、新鲜空气和休闲的最佳选择。

佐鲁中心

Zorlu Center

设计公司：DS Architecture – Landscape

设计团队：Deniz Aslan, Elif Çelik, Saime Selda İpek, Deniz Tümerdem, Zeynep Emektar, Deniz Subaşı, Doğan Onur Araz

地点：Turkey（土耳其）

面积：12 840 m²

摄影：Cemal Emden

The aim of the project is to design Zorlu center, as an alternative urban space for Istanbul. The main feature of the plantation language is to treat the landscape as a succession of Bosphorus grove and yet to perceive it as an integral part of Istanbul. Extroversion was adopted rather than introversion. The green shell which surrounds the structure in third dimension associates with the hill of Istanbul that was once used as the recreation areas of the city. This new built up niche, is adopted as an urban square that will turn out to a new social experience and yet, it is erected as a new kind of nature that will erase the existing idea of shopping mall.

Zorlu Center which is one of the most significant investments located at the heart of business and urban milieu of Istanbul, is a unique project in terms of smooth connections the its surroundings. Zorlu Center is located in Levent / Istanbul a central location, close to the bridge and important districts of the city. As a result of its settlement on possibly the highest profiled site in Istanbul, it has the chance to have an excellent Bosphorus and the city view. Zorlu Center shelters a base structure that is also the substructure of the landscape topography and four towers rising on it. Due to its geometrical surfaces, the landscape roof, so called 'The Shell', is the most significant part of the project. And yet the lost green space on the ground is regained through the landscape design on the shell.

Even though, each component of the project aims carefully to be a part of the

ZORLU
CENTER

ZORLU
CENTER

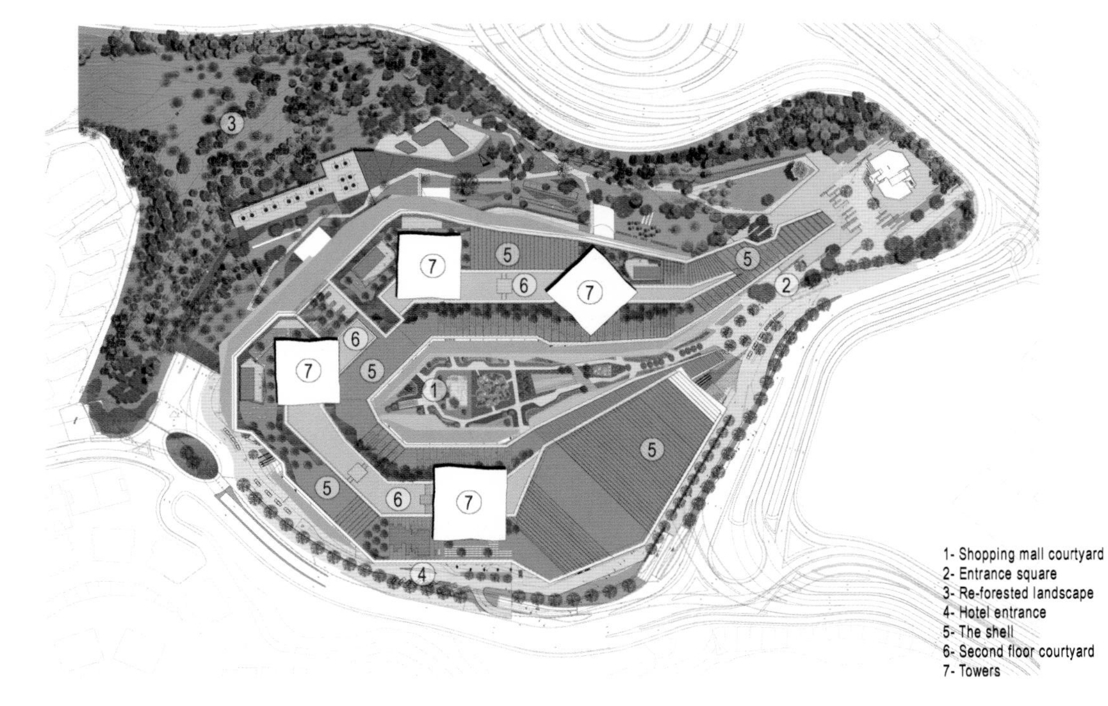

nature, seven main fields come forward as a result.

1. The planting design which contributes both to the landscape of Istanbul and surrounding highways, provides a herbal filter to the drawbacks of those roads.

2. The recreation area functions as a close up garden of the residential area. Places like swimming pool and children' club that are located in this area aim to support the natural environment without deviating the succession of the garden settled in an artificial topography.

3. The square, which functions as a public and a semi-public space, a meeting place for the city, provides a demonstration area in which water plays a major part.

4. The courtyard that allows alternating

functions is the extension of the main square. The well known plant of Istanbul; Pinus pinea is used as the main tree of this space.

5. The facades of the building are treated as the parts of the nature that open into the private gardens of the buildings and the hotel. While these areas are fully designed as a part of nature, the environment that is created allows both water conservation and a new microclimate.

6. The main holistic characteristics of the project are the courtyard and the residential terraces. The plantation islands that reflect the light are located on the thin water surfaces.

7. The valley park might be taken into consideration as a garden collection that is fully themed of Istanbul and as the most valuable and important acquisition of Istanbul.

佐鲁中心是伊斯坦布尔的新地标。其景观设计为简约风格，是连接博斯普鲁斯海峡与城市的纽带。三维结构环绕着绿色的立面，与曾是城市游玩、休憩区的坡地融为一体。这是一个给人们带来全新体验的城市广场，也重新定义了购物中心的概念。

佐鲁中心位于伊斯坦布尔核心商业地带，既卓尔不凡，又与周边环境浑然一体。它地处枢纽，交通便利，可纵览博斯普鲁斯海峡和全市景观。以裙楼为基础，四座塔楼拔地而起。景观屋顶因其几何表面成为整个项目中最抢眼的部分，地面上未能实现的绿地在这里得以实现。

该项目各个组成部分皆道法自然，主要体现在以下七个方面。

1. 种植设计既是城市景观，也是周边高速公路的草本过滤器。

2. 以居住区花园作为休闲区域，建有游泳池和儿童俱乐部，即便在人造地形环境中，也力求自然。

3. 广场既是一个公共空间，又是一个半公共的空间，聚会场所内还建有水景展示区。

4. 多功能庭院是主广场的延伸，以伊斯坦布尔著名的石松为主要绿色植物。

5. 通向私人花园的区域自然天成，其设计有助于节约用水和创建局部小气候。

6. 庭院和露台是景观整体风格的缩影，种植槽位于浅浅的水面上。

7. 谷地公园集中体现了最具伊斯坦布尔特色的风格。

TOM'S KITCHEN
ISTANBUL
TOM'S KITCHEN

joker
SuperKids

TABEVİ
Boğaz

特朗普大厦

Trump Towers

设计公司： DS Architecture – Landscape

设计团队： Deniz Aslan

地点： Turkey（土耳其）

面积： site 23 370 m^2, landscape 13 300 m^2

摄影： Gürkan Akay, Cemal Emden, Günseli Döllük

Trump Towers, one of the most significant investments located at the heart of business and urban life of Istanbul, is a unique project that has a smooth connection with the surroundings. It rests on a total area of approximately 23 370 m^2, 13 300 m^2 of it designed as open space. The design which shapes both the urban and the private use-oriented outdoors and contains the conceptual design of the terrace and the courtyard of the towers, was based on a linear sequential setup that defined the utilization of all the gardens.

The differences between the vegetational and rigid material, which were designed in a linear pattern, both facilitated a 3-D perception of the garden and provided a natural path within the garden. The freshing and elevated feeling created by the two towers was repeated by means of the strong horizontality on the garden planes. The concept of the entire project aimed to simplify the perception of the spaces through a common design perspective.

While designing the entrance square that forms an interface with the urban life, it was planned to create an urban area that attracts people to the building, and also that is used as a transition and a service area. The square became a meeting point with its catering platforms, its pool that matches the refraction of the ground, its green wall, which is also a symbol of eco-technology, that filters out the noise and dirt of outside.

An impressive garden was aimed for in the terraces at the high levels that is far away from the city's chaos. It was intended that the linear lines, on which

özgürce yaşa!
TOWERS

N
1. Drop-off point
2. Cafe
3. Entrance Square
4. Reflection Pool
5. Indoor Swimming Pool
6. Sunbathing terrace
7. Elevation +9,50 Terrace
8. Cafe
9. Tower 1
10. Terraced Garden
11. Elevation +4.90 Terrace
12. Mid-Gardens
13. Walkway
14. Tower 2

the design is based, is to be felt most intensively in this garden, where several utilization possibilities were offered to the user as a multi-purpose open space. While one of the terraces was designed as a transition, walking and recreation area with the highest botanic intensity, the other terrace was designed as a garden carrying out the function of an open space and sunbathing area. The linear water elements in the sunbathing area was included in the design in order to cool the area in the hot weather and to create several reflecting surfaces.

The gardens of the inner spaces of Tower 1 and Tower 2, containing offices and residential areas respectively, are created without irrigation necessities. It was aimed to form a sustainable inner garden and including a small joke in the design. As planting material, imitations with striking colors were preferred and an abstract garden image was created.

特朗普大厦位于伊斯坦布尔核心地带，完美地与周边环境融为一体。其占地面积约达 23 370m²，其中 13 300m² 为开放空间。无论公共区域还是私人空间，均采用了线性设计的方法。

植物和硬质材料在设计中“碰撞”，既增强了花园的立体感，又形成了自然的路径。不同模式的组合，使得场地在水平及垂直方向都具有流动性。设计师借助花园的水平线条，强化了两座塔楼的轻盈和高耸感。整体设计相当简约。

入口广场与城市紧密连接，成为一个缓冲过渡服务区，吸引人们到这里来。此外，作为生态技术的象征，绿墙和多样化水元素的运用隔绝了外部噪声和污染。

裙房上的屋顶花园远离城市喧嚣，直线形的设计为人们提供了多功能的开放空间。一个露台可以休闲、散步，另一个露台则有日光浴花园。线性设计的水景在日光浴花园也有运用，用以降低区域温度，并形成水中倒影景致。

两座塔楼内部是办公区和住宅区，并建有可持续性的内部花园。照明设计同样突出了设计的线性风格。每个功能空间的地板图案、水景设置、绿植元素、灯光效果都个性鲜明、丰富多彩。

ELEVATION +4.90
TERRACE
ENTRANCE SQUARE

TOWERS
SERGİSİ
TRUMP TOWERS MALL
AV MI? AVCI MI?
TRUMP TOWERS MALL

都柏林
天然气公司

Bord Gais Dublin

设计公司：TOPOTEK 1

地点：Ireland（爱尔兰）

面积：20 000 m^2

摄影：Hanns Joosten

The site of the new Bord Gáis Network Services Centre is located on a triangular parcel close proximity to the M50 motorway, which forms its northern border. In response to its location between urban development and existing green landscape and with regard to the existing site conditions, the landscape strategy seeks to integrate these aspects by setting the required car parking and loading areas in a new park-like landscape that provides environmental, aesthetic, and microclimatic amenities to staff and visitors. A new common entrance plaza for the existing building and new Services Centre is distinguished by a large group of oaks and an attractive new water feature. Terraced parking draws inspiration from traditional Irish hedgerows, and features a central infiltration area with wet meadows and landscaped pedestrian circulation. The treatment of stormwater run-off occurs throughout the site in different forms, from the wet meadow in the car park to a swale along the eastern border, to a water feature with attenuation pond at a newly unified entrance to the site. An Integrated sustainable design approach combines microclimate, landscape, transport and a compact building volume with low energy design to establish a service facility that humanizes and civilizes the environment of the corporate workplace. With its protective skin, compact plan and interspersed gardens the building provides acoustic shelter from the nearby M50 motorway, improves the site's microclimate and establishes a high quality, highly adaptable and permeable work environment.

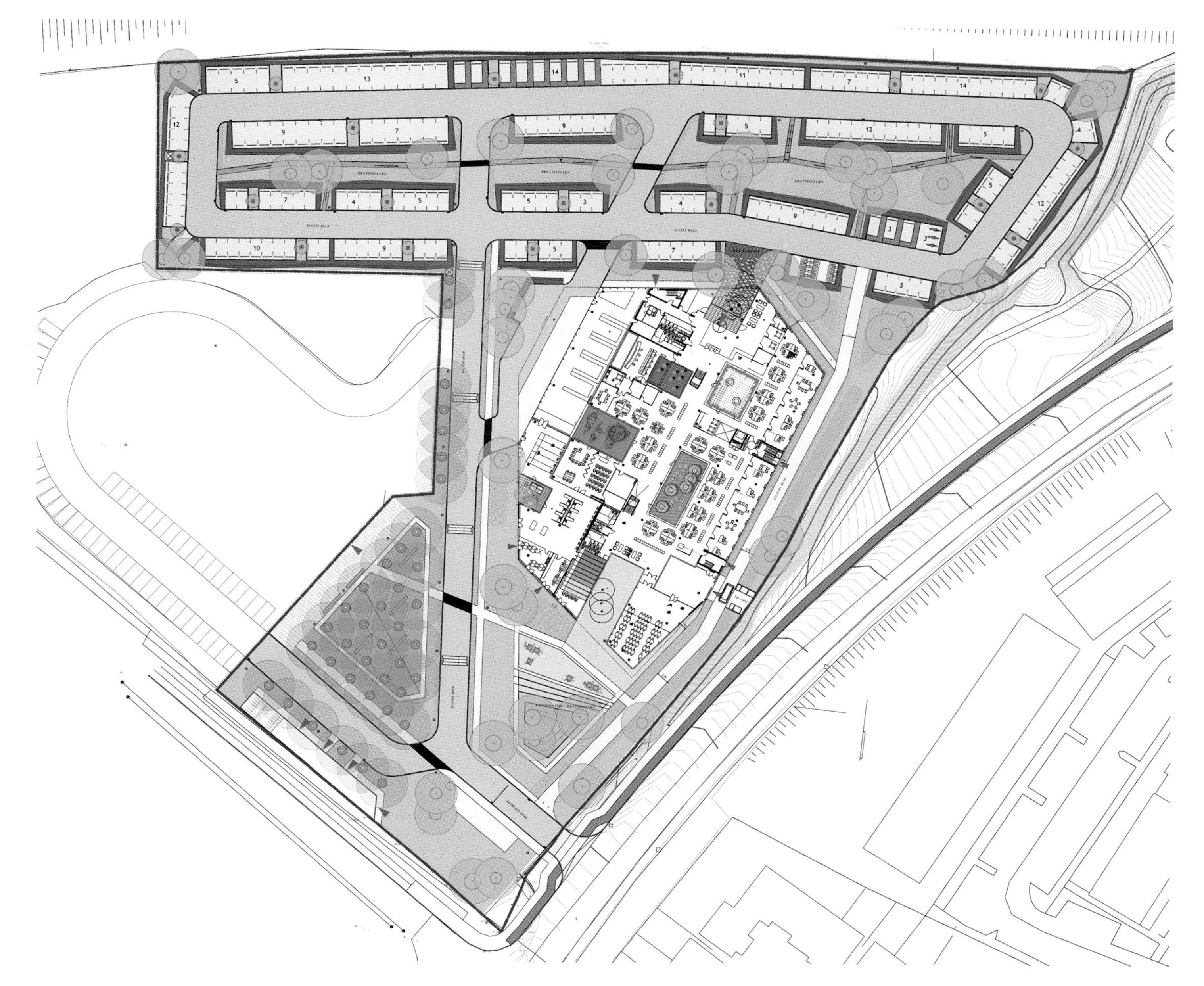

新落成的都柏林天然气公司网络服务中心位于高速公路以南的一块三角地上。基于城市发展和现有绿化及其地形条件，景观设计目标力图整合所有既有元素，使停车场和装卸区犹如公园般环保、美观，让园区设施为员工和访客营造怡人的环境。现有建筑和新服务中心新的入口广场橡树成林，水景迷人。梯田停车场的灵感来自传统的爱尔兰树篱，中心渗透区覆盖着湿草甸，园区内行人通道穿行其间。区域内采取不同形式的雨水径流处理方法，如从停车场到沿东部边境的洼地均铺设了湿草甸，新建的入口建有水景。气候、景观、交通和紧凑的建筑体量相结合，采用低耗能设计，达到综合可持续发展的目标，使得企业工作环境服务设施更加人性化。通过外墙、园林等的运用，减少了高速公路带来的噪声干扰，改善了园区微气候，营造出品质一流、适应性强的工作环境。

威悉河商业空间

Weserquartier

设计公司： Rainer Schmidt Landschaftsarchitekten GmbH

地点： Germany（德国）

面积： 24191 m^2

摄影： Rainer Schmidt Landschaftsarchitekten, Besco, H. Siedentopf

The Weserquartier, situated at the entrance of Überseestadt district and within walking distance of Bremen's city centre, extends along the Weser River and the riverside promenade.

Its special location by the riverside and bordering the former ramparts of Bremen's historic centre makes the Weserquartier the gate of Überseestadt. A hotel, office blocks, and a theatre with a casino were built as the first phase, followed by the "lighthouse" of the quarter – the Weser Tower – in 2010. Planned by Murphy / Jahn Architects, Chicago, the tower is 82 metres tall and acts as a landmark.

The landscape architecture was to create the common artistic ground for the heterogeneous architecture on site and help to form the quarter's identity.

Like a carpet the landscape design extends homogeneously over the entire plot. The basic geometrical grid in the form of a continuous linear pattern interconnects the individual buildings, their typologies and uses. It enhances the tangibility of the entire quarter in its urban context. The paved surfaces, finished with natural stone, and the paths between the architectural grass fields follow a coherent layout, which promotes the readability of the exterior spaces. The subspaces are accentuated and ordered by contrasting light-coloured granite and dark basalt paving stones. Gravelled paths between the planting beds perpetuate the dominant linear structure of the pavement. A sequence of daffodils, foxtail lilies and Eulalia grass (Miscanthus) defines the colours and the scent of the planted areas.

At the banks of the Weser, stepped terraces are the perfect place to bask in the sun. However, the terraces can also be used as lounges by the adjoining restaurants and cafés or as seats for spectators during events.

威悉河商业空间坐落于 Überseestadt 街区的入口处，临近不来梅市的市中心，沿着威悉河沿岸延伸。

本项目沿河的特殊地理位置及与不来梅历史中心紧邻的位置，使其成为该街区重要的商业区。一期建筑有酒店、办公大楼、剧院和赌场。二期建筑是于 2010 年建成的威悉塔，作为地标建筑，该塔有 82m 高。

建筑群内各种各样的建筑周围都使用了相同的地面铺装，这因此成为该区的标志。

像地毯一样的景观设计延伸至整个商业区。设计师采用几何网格的连续线性图案的形式，将单独的建筑物相互连接起来，增强了片区建筑的整体性。地面的铺设、天然的装饰和石材及草地之间的道路，形成一个连贯的布局，使外部空间更具可读性。设计师使用了对比强烈的浅色花岗岩和玄武岩黑铺路石。种植区之间的砾石铺装延续了主要地面铺装的形式和风格。种植园内种植了水仙、独尾草和中国芒，色彩艳丽，芳香宜人。

威悉河畔两边的阶梯式露台是晒太阳绝佳的场所。然而，露台也可以用来作为隔壁餐馆和咖啡馆的休息室，或在活动期可摆放座位。

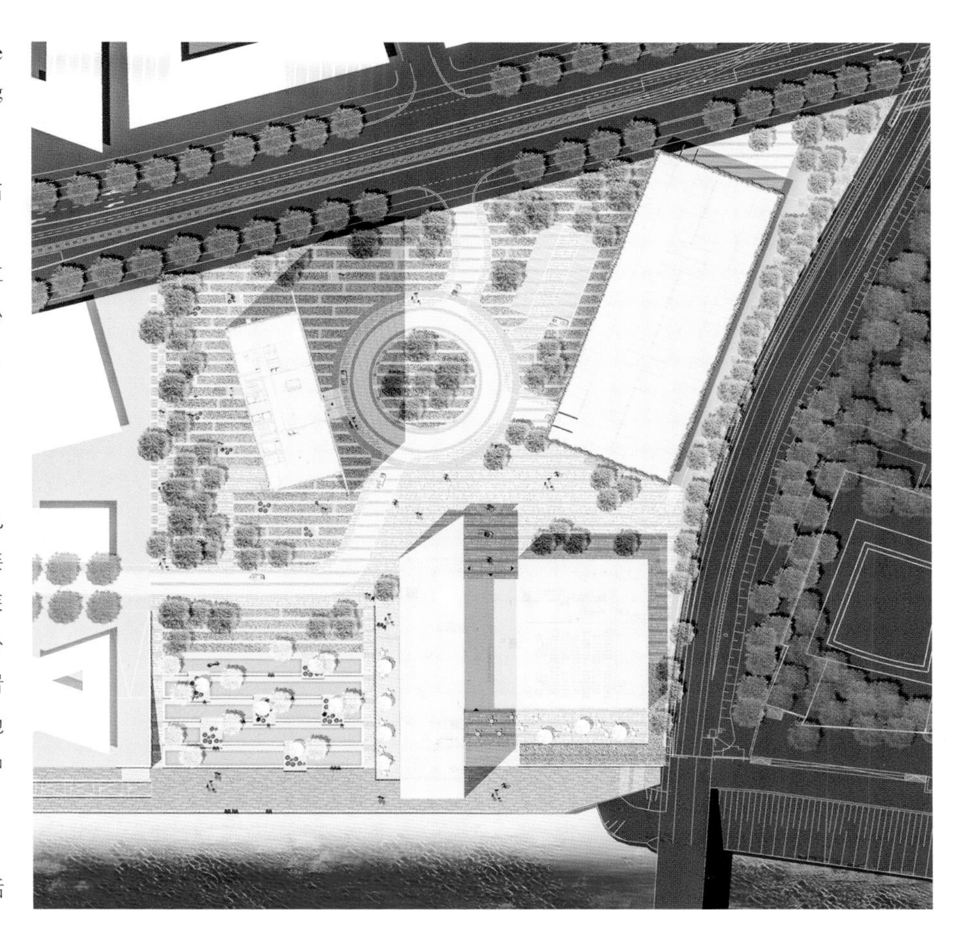

sichtbar

冠利大厦

Crest Prime Tower Shiba

设计公司： studio on site (Landscape Deign), MHS Planners, Architects & Engineers (Architecture Design)

设计师： Chisa Toda, Toru Mitani

地点： Japan（日本）

面积： 4874 m²

摄影： studio on site, SS Tokyo

The Crest Prime Tower Shiba is a 39 th story skyscraper housing in central Tokyo. It is constructed next to the intersection of the Yamanote line and Tokyo metropolitan expressway, which is not a confortable neighbors for housings.

The main landscape design idea was to create tree foliage canopy, surrounding the building, which provide shades and cozy human scale spaces for people.

The building has three gardens. The west side of the building is a Greeting Garden, where is the main entrance of the building. This is the most formal garden with evergreen tree rows and trimmed evergreen bushes. The south part is called Forest Garden, where has Lagersroemia indica and other deciduous trees. The east part garden is called Seasons Square. There are bamboo bushes, playing lawns, and flowers at this area, where kids can play safe.

Under the metropolitan expressway, there is a canal connected to Tokyo bay. Traditional Japanese leisure boat called Yakatabune Station is at the canal. To separate the private housing garden area and the public Yakatabune Station, the designed light wall was constructed. The wall also function as a wind break for the canal.

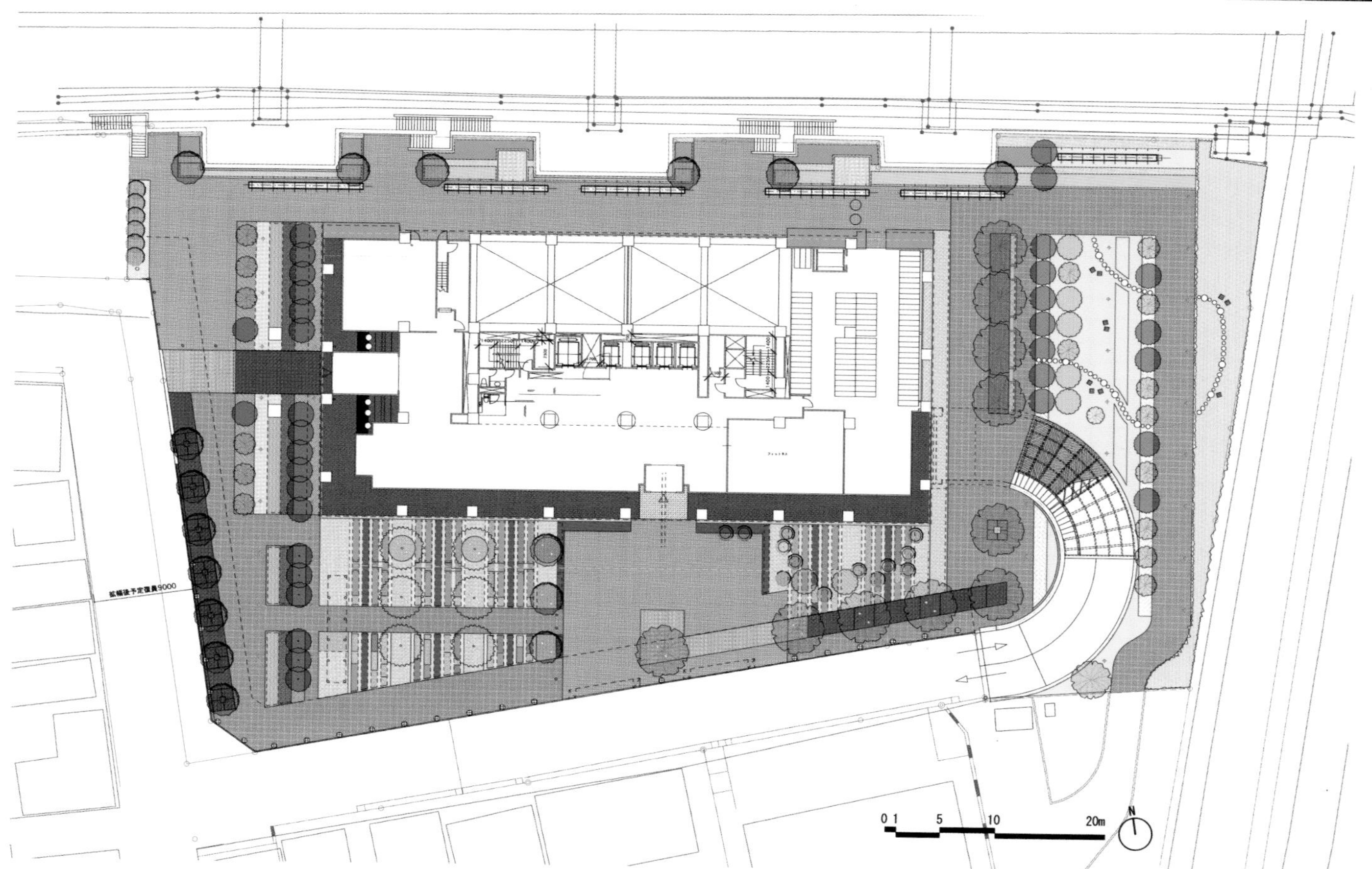
0 1
5
10
20m
N

冠利大厦是东京市中心一座 39 层高的摩天大楼，位于山手线和东京都市高速公路的交叉处。

其景观设计主要利用植物包围建筑，为人们提供阴凉、舒适的人性化空间。大厦有三座花园。西侧是迎宾花园，也是建筑的主入口。这是最正式的花园，配有常绿树列及修剪整齐的常绿灌木。南侧是森林公园，种有落叶乔木。东侧花园被称为四季广场，有竹林、活动草坪和鲜花，孩子们可以在这里安全地玩耍。在都市高速公路之下，还有一条运河连接到东京湾。河上，有传统的日式休闲船运行。为了分隔私人住宅花园和公共船运，特别设计修建了轻质墙，此墙亦可起到防风作用。

保利国际广场

Poly International Plaza

设计公司： SWA

项目团队： Skidmore Owings & Merrill (Architect), Guangzhou Landscape Planning and Design Institute (Associate Architect), Flack & Kurtz, CMS Collaborative, Inc. (Fountain Consultants)

地点： Guangzhou, China（中国广州）

面积： site 57 hm^2，landscape 11.6 hm^2

摄影： Hanns Joosten

Poly International Plaza is an innovative office and exhibition center development located in China's Guangzhou trade district. Sited along the Pearl River and adjacent to historic Pazhou Temple Park, the project presents a precedent toward integrating development with its site and context, embracing the place of sustainability in the society's rapid move toward modernization. Located in Guangzhou, China, the 57-hm^2 site is part of the city's new exhibition and trade district. The property was formerly agricultural fields with water channels linked directly to the Pearl River. Consisting of two slender north/south facing wafer towers coupled with low-rise podium buildings, the architecture is diagonally offset around a large central garden court. This building configuration and massing capture the prevailing breezes, channeling them through the central garden. As part of a broad architectural plinth, the entire garden court is elevated 1.5-meter to enhance the effects of these breezes. To instill a cooling effect, water surfaces are strategically configured to engage these breezes. During monsoon season, the watercourse serves to partially store and convey storm runoff across the site. Substantial building programs located below grade translate into more green space onsite. Approximately one-third of the site landscape is developed as roof gardens. To accommodate anticipated pedestrian usage, dense tree plantations establish a significant tree canopy that shades much of the ground plane, aggressively reducing the heat island effect. To support and sustain the tree canopy, a 1.5-meter

租售中心
lease

soil layer is provided over much of the structures. Broad allees of trees establish a shady frame around the architecture and effectively shade the west and east facades of the lower plinth buildings. The elements designed by SWA create a new, sustainable office and exhibition environment for Poly International Plaza.

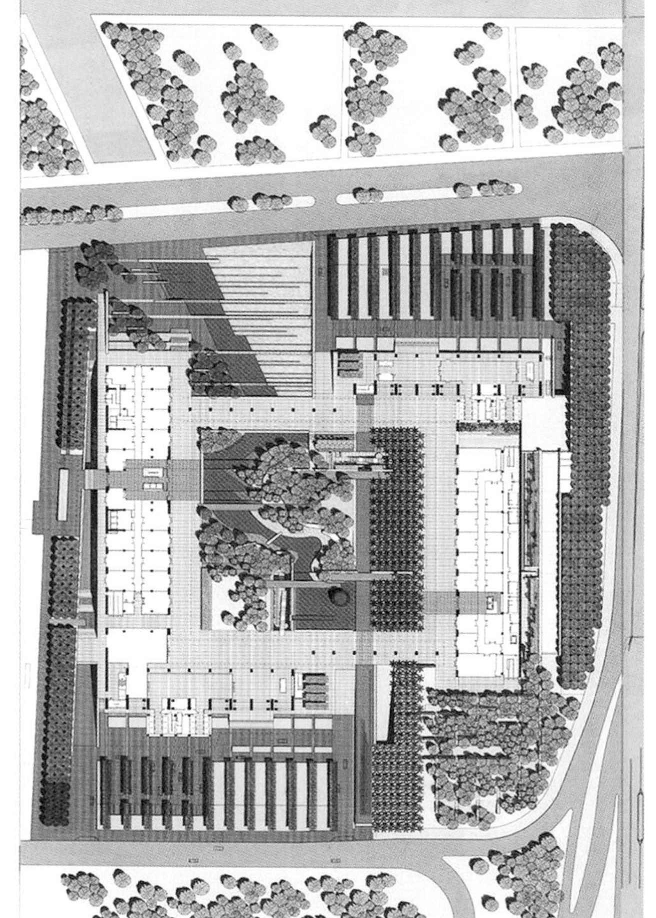

保利国际广场坐落于广州贸易区内，是一个以办公和商贸展览为主的创新型综合开发项目。项目设施沿珠江滨水地带而建，毗邻历史悠久的琶洲塔公园。因其所处的独特城市地理环境，项目有关各方提出颇具远见的规划设想：将项目开发与当地的城区发展融于一体，以可持续发展为核心，与社会整体的高速现代化发展方向保持一致。项目总占地为 57hm^2，是广州市新建展览与贸易区的一部分。项目场地原先是一片农田，其灌溉水渠直通珠江水系。场地中央花园对角线的两端，两座合晶体塔式建筑拔地而起，底层周边是它们各自的附属裙楼。这样巧妙的建筑布局，有利于将海风引入中央花园。作为大规模建筑基座的一部分，中央花园整体向上挑高 1.5m，进一步有效地促进海风的引入。为了更好地达到项目场地的降温效果，设计师们策略性地设置了大面积的地表水景，以配合海风对广场建筑及周边区域进行降温处理。在季风季节里，这些水景设施可部分存储相应的雨水径流，并将其输送到广场的各个角落。项目场地以坚实的地下结构为基础，营建了大规模的绿地空间，此外，占地约 1/3 的裙楼建筑几乎都建有屋顶花园。树木密集地种植于广场空间内，形成庞大的天然步道遮荫设施，有效缓解了热岛效应，为行人们营造出惬意、凉爽的步行环境。种植区内铺设 1.5m 厚的土壤层，以支撑和维持这些树木的生长。大规模的林荫步道成了环绕建筑物周边的林下空间，有效地为东西朝向的低层裙楼建筑抵挡了烈日的侵袭。SWA 各种巧妙的景观设计元素为保利国际广场创建了一处全新的可持续的办公和展览环境。

成都赛门铁克

Symantec Chengdu Campus

设计公司：SWA

地点：Chengdu, China（中国成都）

面积：$1hm^2$

SWA provided landscape design for Symantec's new research and development complex in Chengdu, China. The site was previously inactive and banal, with SWA's design reinvigorating the area, linking the building program and connecting the site to the larger city. The landscape design creates a "brocade", weaving together the building and site program, and creating an oasis amidst the dense, urban location. The design takes into account the city of Chengdu by paying attention to key infrastructural elements. In order to mitigate storm water runoff from the site, the design incorporates an extensive water filtration garden, which provides a natural solution to this common infrastructural problem. An intricately programmed roof garden adds to the connection of building and site to SWA's design, creating functional and environmentally sensitive urban outdoor spaces.

设计公司 SWA 为赛门铁克在中国成都新建的研究和开发综合楼提供景观设计。该区域原本松散无奇，SWA 的设计不但让建筑间建立了有效联系，更将整个区域与城市连接，使该地块整体环境得以提升。景观设计如同一匹“锦缎”，将建筑与区块编织为一体，在密集的城市中开辟出一片绿洲。设计充分考虑了成都的地域特点。为了减少雨水径流，特别设计了大面积的雨水过滤花园，为排水提供了一个自然解决方案。楼顶建有错综复杂的屋顶花园，打造了一个功能齐全、环境优美的城市户外空间。

海洋金融中心

Ocean Financial Center

设计公司： TIERRA DESIGN (S)

地点： Singapore（新加坡）

面积： 6 109 m²

摄影： Amir Sultan

The tropical gardens in the sky at the Ocean Financial Centre were designed within the interstitial spaces between the building structure and façade framing. Creating a connection to the natural environment, the vertical landscaping at the highest floors from the 39th to the 43rd level provides a lush setting. These areas provide much needed relaxation and relief to users who spend most of their time in the artificially ventilated environment of an office building. In response to the hard surface of the structural framing of the building the landscaping provides a contrasting effect softening the lines and creating an intimate experience. The canopy trees at the 41st floor create an interesting play of scale and different textures and colours of planting offer a vibrant visual treat. At another level, landscaping frames 2-storey high columns at 41st floor and 1-storey high columns at 42nd and 43rd floors creating a view of vertical green planes for the office space and framing the dynamic city skyline beyond. The vertical green frames are comprised of planters located at every 1.5m. The lightweight mesh structure is planted with fast growing Thunbergia grandiflora climber species to ensure a full green cover. The showy purple flowers and the large leaf foliage add to the tropical feel of the interactive garden space. A combination of profuse flowering plants and colourful textured leaf foliage of shrubs at the base of the green columns lend an exciting and vibrant feel to this limited landscape space. At the 43rd level,

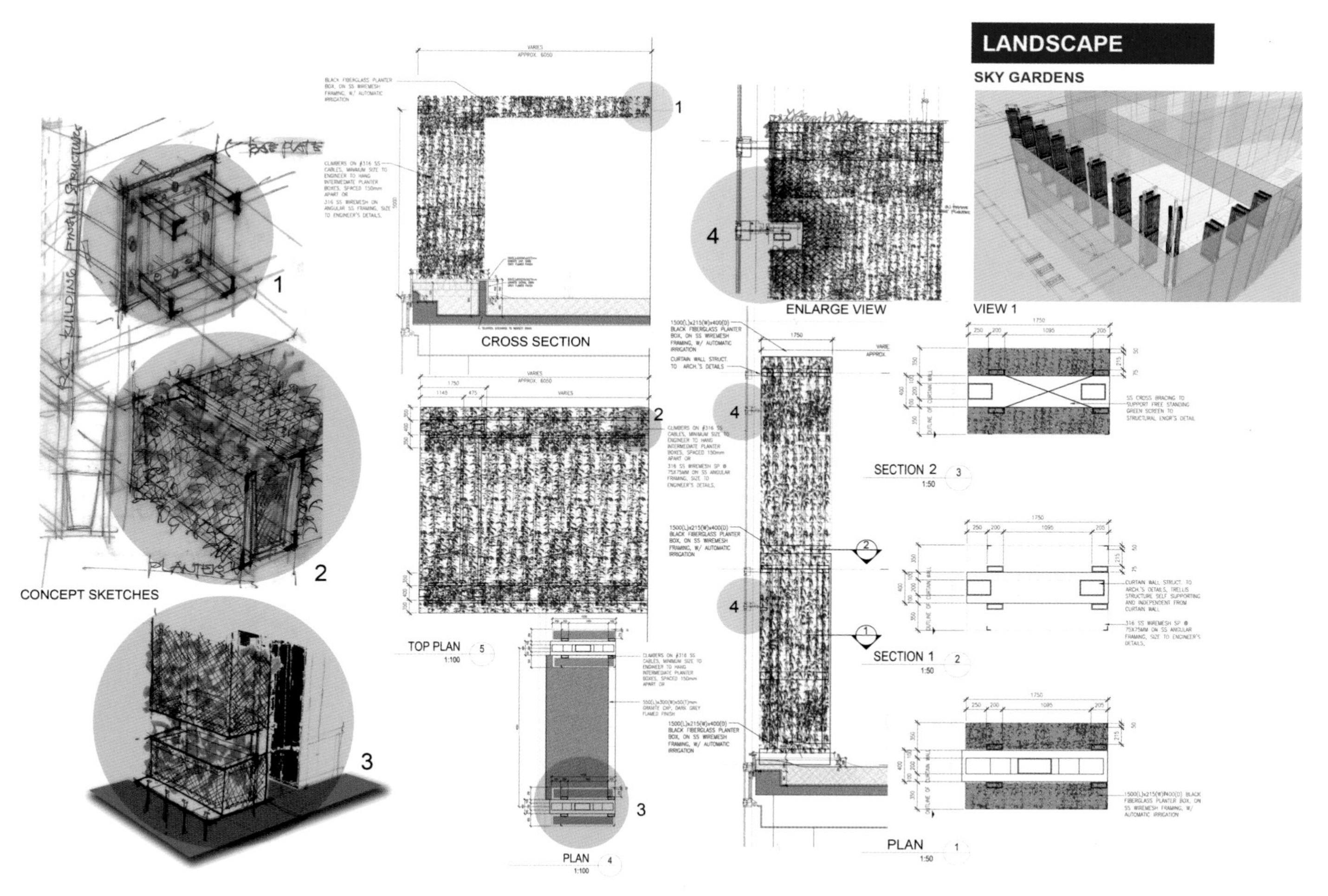
LANDSCAPE
SKY GARDENS
CONCEPT SKETCHES
CROSS SECTION
TOP PLAN
1:100
PLAN
1:100
ENLARGE VIEW
VIEW 1
SECTION 2
1:50
SECTION 1
1:50
PLAN
1:50

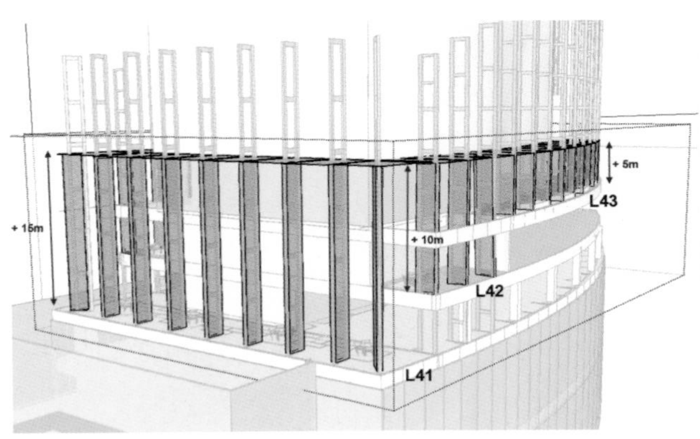

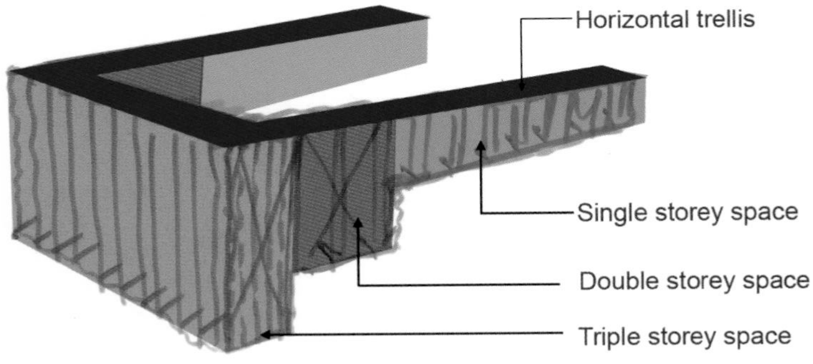

the green columns are combined with horizontal trellis feature to create a green arbor in the sky. Horizontal planter trays at regular intervals planted with trailing plants to cover the trellis form a soft green canopy for the sky garden users. Auto-irrigation & fertilisation system installed for all the planter areas ensure that through timer controlled drip nozzles, slow, steady and precise amounts of water and nutrients are provided. With slow application rate, water seeps into the soil media and is fully absorbed by the capillary action of the root system. Rain sensors are installed at planters to cut off irrigation during rainy days. All these measures ensure that water is judiciously used in just the required amounts for the plants to thrive and sustain the lush urban high-rise gardens at the Ocean Financial Centre. A series of tropical garden views, outdoor walks and seating at different levels create a work environment that is refreshing and unique. The landscaping within the interstitial spaces between the building structure and the façade framing blurs the boundary between the built and the natural environment.

海洋金融中心的空中热带花园处于建筑结构和外观框架的夹缝中。最顶部 39~43 层的垂直景观创造了一片郁郁葱葱的小天地，为那些整日待在人工通风环境中的写字楼用户提供了放松之处。景观软化了结构框架冷硬的表面，带来丝丝亲切感。41 层上的大树树冠遮天蔽日，形形色色的植物带来充满活力的视觉享受。不同楼层不同高度的景观框架搭建出办公空间的垂直绿化面，丰富了城市的天际线。垂直绿化框是由间隔 1.5m 放置的花盆组成。轻便的网状结构，种植着快速生长的藤蔓植物。

在有限的空间里，多姿多彩的开花植物和阔叶灌木错落有致，突显了交互式园林空间的热带风情。在 43 层，绿柱与网格纵横交错，搭出一个空中绿色凉棚。所有种植区域都安装了自动灌溉及施肥系统，通过定时器控制滴灌喷头，缓慢、稳定并精确地提供水和营养物质。水慢慢渗透到土壤中，通过植物根系的毛细作用被充分吸收。阴雨天，雨传感器会自动切断灌溉系统。这些措施确保了供水恰如其分，植物茁壮成长，高层花园郁郁葱葱。各楼层的花园便于用户散步和休息，营造出清新独特的工作环境。这一存在于建筑结构和立面框架夹缝中的景观模糊了建筑与自然的边界。

塞西尔街 158 号

158 Cecil Street

设计公司 : TIERRA DESIGN (S)

地点 : Singapore（新加坡）

The site is an existing 10-storey non-descript, even boring office building located at 158 cecil street in singapore with naturally ventilated façades covering an 'external atrium' space. Three storeys were to the building some years ago extending beyond the existing building elevation. New columns and transfer truss were added over the building, thus creating this interior naturally vented 9-storey atrium volume.

The approach to greening the interior atrium space was firstly to identify the opportunities which would allow for a methodological, systematic and modular method of placing, as large as possible, planes of green vertical surfaces for planting "beds". Secondly, it was necessary to ensure ease of maintenance and accessibility. Thirdly, plants had to have sufficient amounts of water and enough light to be sustained indefinitely. Since the atrium is east facing and its opposite neighbors high rise buildings, with cecil road in between, natural lighting would be insufficient to sustain the planting as the subsequent vertical gardens are totally in its own shade. There would be also no direct sunlight to speak of. In regard to the flat planes of planting "beds", it was fortunate that upon closer inspection, there was sufficient flat walls and shear wall columns which could be clad with the plant system.

Thus during the arduous task of preparing these surfaces for planting, physical accessibility for maintenance had to be planned, a water system

for irrigation incorporated and artificial lighting to supply a minimum 1000 lumens on all planting surfaces provided for.

项目位于新加坡的塞西尔街 158 号，是一个 10 层高的无装饰的办公建筑。建筑拥有自然通风的立面，中间有一个中庭。顶上的 3 层是几年前扩建的。建筑增加了新的柱子，建造了一个新的中庭内部空间。

中庭的绿化首先要确认设计方案，设计师计划采用系统的模型方法，尽可能地种植出垂直绿色“外墙”。其次，外墙必须易于保养。最后，植物要有足够的水源和充沛的阳光以维持生命。因为中庭面朝东方，背面是高建筑，中间是街道，因此自然光照不足，背面的植物全部处于较阴暗的环境中。通过近距离地仔细观察，设计师欣喜地发现建筑有足够多的墙面和立柱用于建造植物“外墙”。

在准备植物“外墙”的艰巨过程中，设计师还需要加上维护植物健康的通道、灌溉植物用的水系统和人工照明系统。

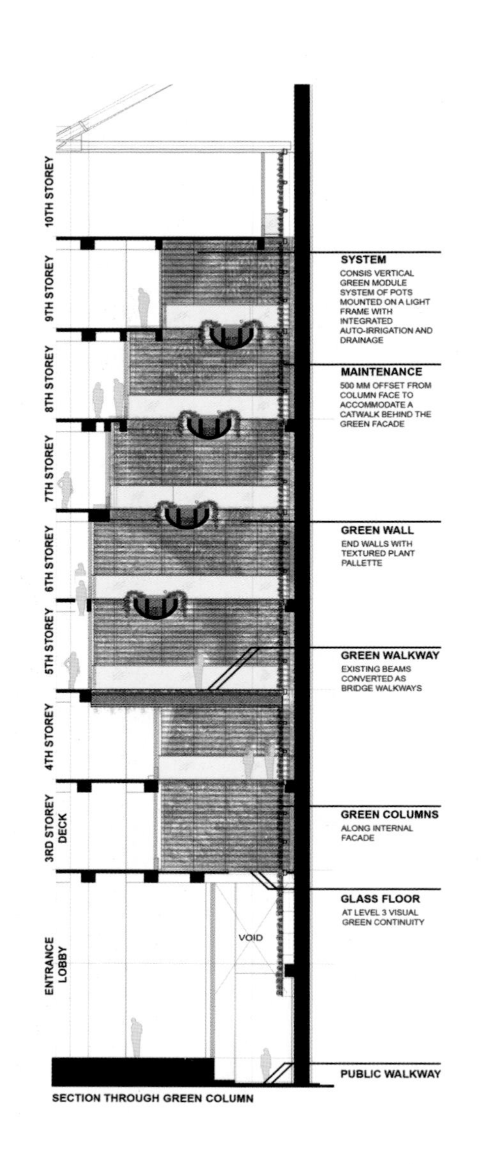

SECTION THROUGH GREEN COLUMN

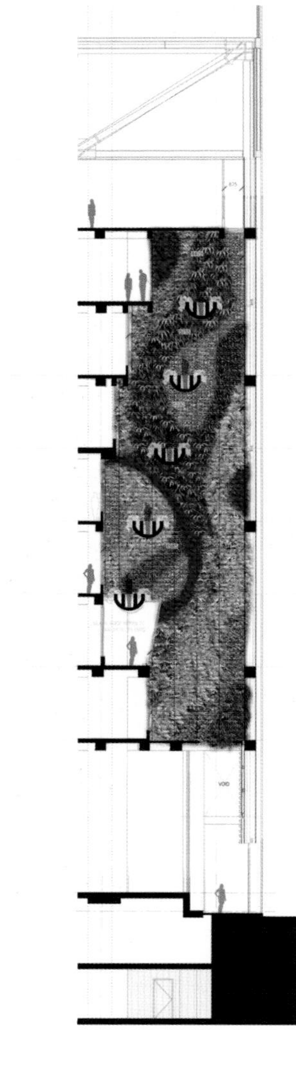

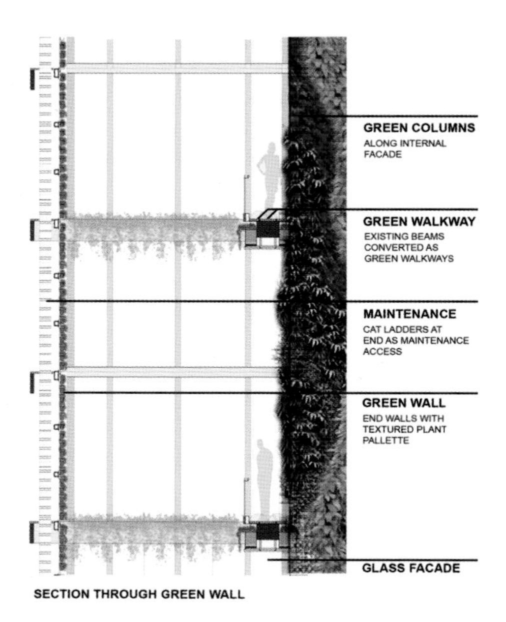

SECTION THROUGH GREEN WALL

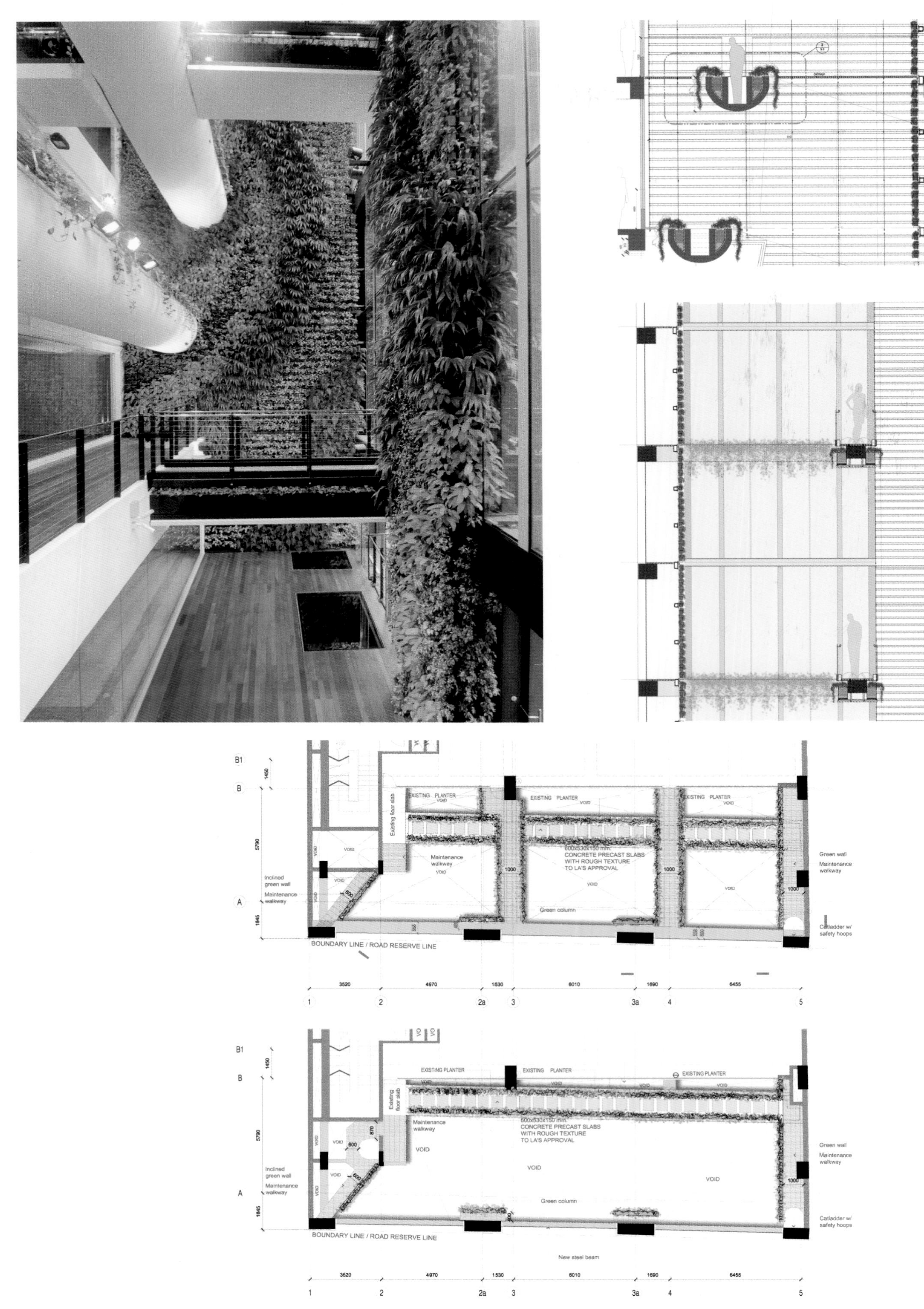

B1
B
A
1450
5790
1845
EXISTING PLANTER
VOID
Existing floor slab
Maintenance walkway
600x530x150 mm.
CONCRETE PRECAST SLABS
WITH ROUGH TEXTURE
TO LA'S APPROVAL
1000
Inclined green wall
Maintenance walkway
Green column
Green wall
Maintenance walkway
Catladder w/ safety hoops
BOUNDARY LINE / ROAD RESERVE LINE
3520
4970
1530
6010
1690
6455
1
2
2a
3
3a
4
5
B1
B
A
1450
5790
1845
EXISTING PLANTER
Existing floor slab
Maintenance walkway
600x530x150 mm.
CONCRETE PRECAST SLABS
WITH ROUGH TEXTURE
TO LA'S APPROVAL
VOID
Inclined green wall
Maintenance walkway
Green column
Green wall
Maintenance walkway
Catladder w/ safety hoops
BOUNDARY LINE / ROAD RESERVE LINE
New steel beam
3520
4970
1530
6010
1690
6455
1
2
2a
3
3a
4
5

德国汉莎航空中心

Lufthansa Aviation Center

设计公司： MK2 International Landscape Architects (in cooperation with WKM, Ingenhoven Architects)

地点： Germany（德国）

面积： 36 000 m^2

摄影： Pieter Kers, HOSPER

The comb like building plan with 10 wings encloses landscaped gardens as buffer zones insulating the building against pollution, emissions and noise. Plants chosen from five continents symbolizes Lufthansa's global connections. All 1,850 office work places have views into the glass roofed gardens and can be naturally ventilated. All gardens are accessible to relax in, or even to organize a small conference.

建筑拥有 10 个外形类似飞机机翼的封闭式的景观花园，花园是一个缓冲区，隔绝掉了外部污染、辐射和噪声。来自五大洲的植物象征着汉莎航空全球化的特征。所有 1850 个办公室都可以欣赏到玻璃屋顶花园的景色，并且感受到舒适的自然通风的环境。每个花园可以用来放松心情或充当一个小型会议的组织地点。

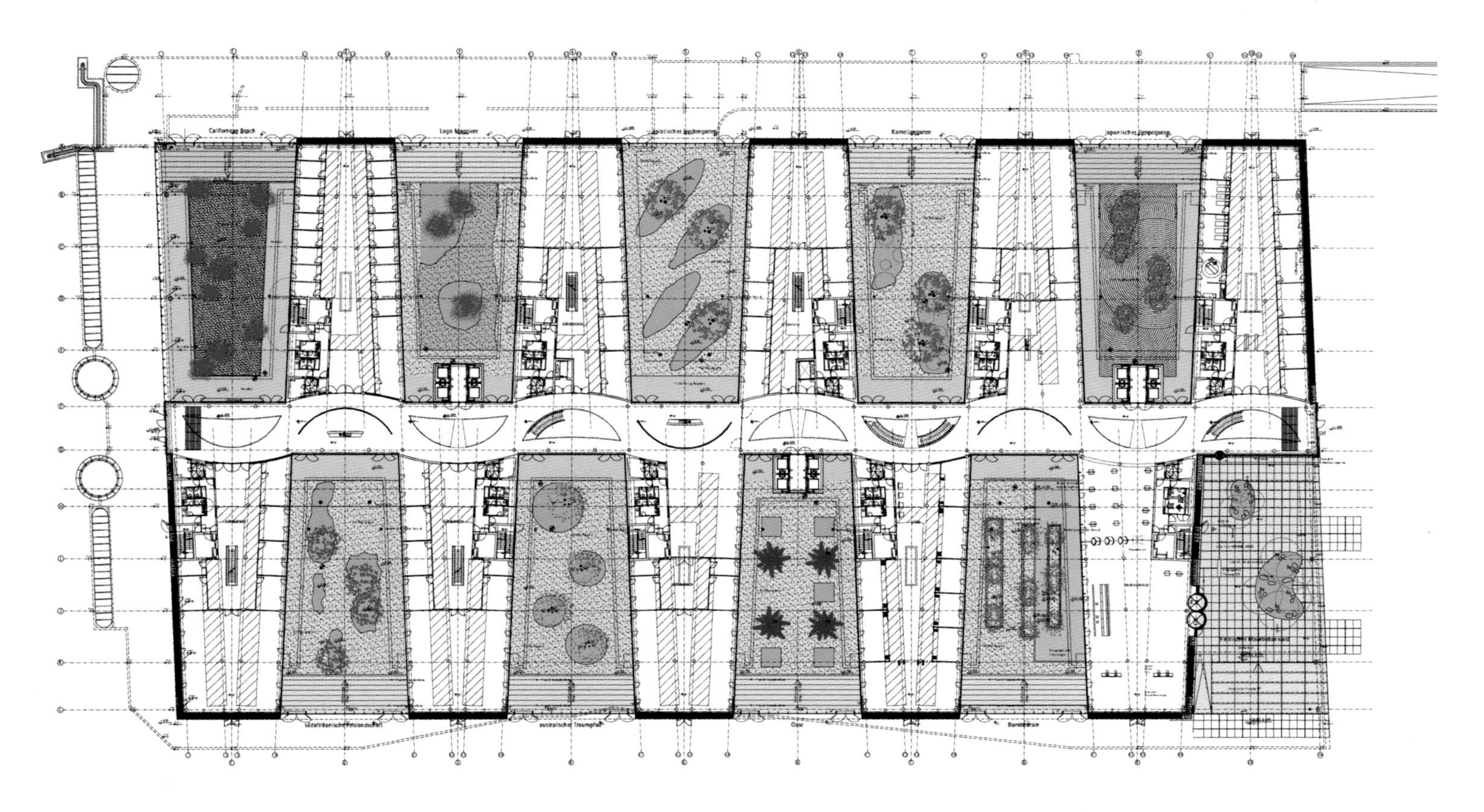
Lago Maggiore
südafrikanische Felslandschaft
australischer Traumpfad
Oase

SAP 创意中心

Innovation Center of SAP

设计公司：TOPOTEK 1

地点：Germany（德国）

面积：5 650 m²

摄影：Hanns Joosten

The garden campus is sited at the Development and Innovation Center of SAP SE on the Jungfernsee in Potsdam. The landscape design responds to the campus architecture while embracing the cultural landscape of historic Potsdam's sweeping vistas, idyllic lakes and manicured castle gardens. The design thus assumes architectural personality combined with strong horticultural and botanic character.

Like a symphony composed as a singular motif comprised of a collection of unique musical elements, the tree assemblage constitutes a singular genus, the Acer, which is celebrated in its diversity. Eighteen different Acers in a variety of form, height and character are composed on site. In autumn one can find a scattering of leaves each with the identifiable Acer shape. The Acer compilation represents local species and others from around the world, reflecting the global character of the SAP SE company.

A large Acer assemblage combined with grass and other herbaceous ground cover accentuates the entrance to the Center. On the eastern edge of the site between the ground level elevation and the lower level of the Center, a cascade of stepped terraces, punctuated by small gardens, connect the two spaces. The stratified stair form of varied dimension, interspersed with Acer species, generates space for a variety of uses.

Outside the cafeteria, on the north side of the site, a hard surface patio is provided for outdoor seating. A large circular bench with an interior fireplace sits beyond in the open grassy knoll dotted with Acer species. Along the site

edge towards the Jungfernsee, a banister is provided for balcony-esque views out to the lake. The overall design is complimented by an extensive green roof, which blankets the Development and Innovation Center's rooftop.

As a large-scale gesture, the Acer tree collection gives unity and defines the horticultural character of the Center. Combined with a mixture of hard and soft scapes, the site provides versatility while offering aesthetic intrigue.

本项目位于波茨坦初航湖的 SAP 发展及创新中心内。景观设计体现出历史名城波茨坦的辽阔的景色，田园诗般的湖泊和修剪整齐的城堡花园。设计具有鲜明的个性和植物种植特色。

设计像是在演奏一首由各种乐器组成的交响乐。一丛丛的树木是整首乐曲中的音符，独立而丰富。18 棵品种不同的枫树形态各异。秋季到了，树叶纷纷落下，随风飘落。每片叶子的光滑度、锯齿形状、宽度、色度都不一样。这里种植着来自世界各地的植物种类，体现了 SAP 创新中心的世界性。

枫树、其他草本植物都种植在中心入口处。在项目的东边边缘地带的低台区有一小块梯田花园。层叠的楼梯形成不同高度的空间，其中点缀着枫树。台地地形较低，装饰着天然石材马赛克。

咖啡屋外边是一块露天台地，为人们提供户外用餐的场地。草坪上有一条环状座椅，座椅旁边点缀着不同品种的枫树。露台紧邻的是美丽的初航湖。人们站在露台上可以欣赏湖的美景。绿色的屋顶弥补了整体设计的不足。

作为一个大型标志，枫树树丛也是 SAP 中心的园艺特征。景观整体用软、硬装饰结合，体现了多样的美学特征。

花园式购物中心水景

Water Feature for the Gardens Mall

设计公司：Aquatic Paradise SdnBhd (Pond Specialist Company)

地点：Malaysia（马来西亚）

面积：30m²

摄影：Ken Lim

Spring, water view, vegetation, lights are perfectly matching with each other. The whole become the landscape focus of the Gardens Mall. The water pool has spring in it, adding visual experience to the customers' sight. The edge of the water pool is made from marble, the customers can have a rest or enjoy the cool by sitting beside the water pool. The central column of the water pool is decorated with various inner common plants. This is a place allow customers to feel the natural smell.

喷泉、水景、植物、灯具完美搭配在一起，成为购物中心的景观焦点。水池中有喷泉，为顾客增加了视觉体验。水池的边缘由大理石砌成，顾客们可以再坐在水池边休憩、纳凉。水池的中央圆柱装饰了各种各样的室内常用植物。这是一处让顾客们感受到自然气息的景观空间。

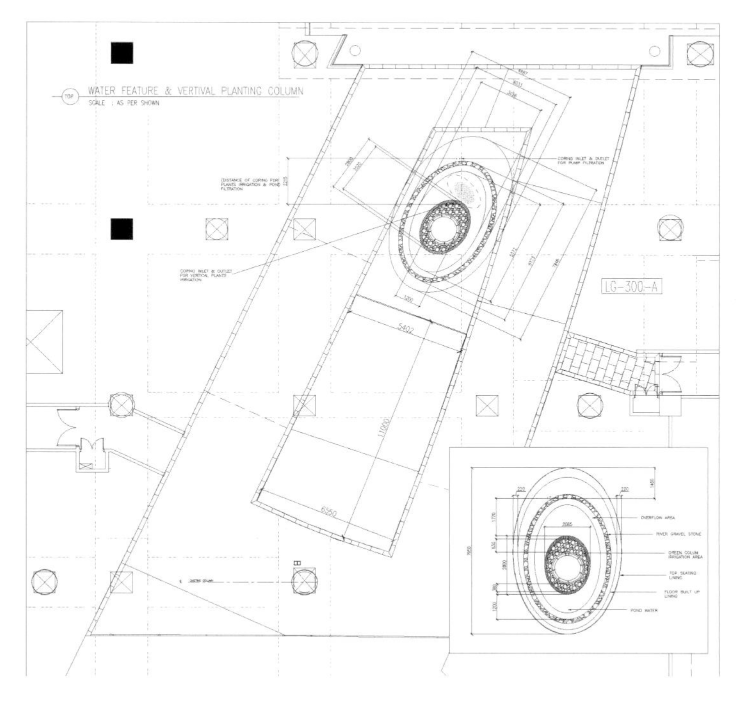
TOP
WATER FEATURE & VERTIVAL PLANTING COLUMN
SCALE : AS PER SHOWN
LG-300-A

空中食宿办公总部后花园

Airbnb Headquarters Courtyard

设计公司：Meyer + Silberberg Land Architects

地点：USA（美国）

摄影：Drew Kelly

Meyer + Silberberg teamed with Gensler and SKS to reinvigorate Airbnb Headquarters in San Francisco's SOMA district. The design creates a flexible social space that fosters new ways of working and invites collaboration.

The courtyard at 888 Brannan Street is conceived to address the changing work habits of San Francisco's tech workers and illuminates the important role landscape, at any scale, can serve in a rapidly expanding city. An existing loading dock was converted into an elevated plaza, connecting the Jewelry Center to the new headquarters for Airbnb. This unsightly sliver of space was radically reconfigured to bring a striking landscape amenity to the complex.

Built around an imported specimen maple tree, a courtyard was constructed by suspending wood decking over the depressed loading dock. A ring of 'semi-precious stones' adorn the tree, distinguishing the courtyard, and providing a sheltered social space for tenants to work, eat lunch or socialize. Other design components include a wall of succulents concealing a mechanical corridor, and a wood-clad wall that when lit, magically transforms the space at night. The design of the courtyard provides multiple configurations for congregating and reflects a work culture that encourages informality and flexibility.

A significant design challenge was how to treat the deep, existing overhang

of the Jewelry Center which created a shaded and oppressive space. The design solution looked to bring elegant detail to the surfaces and artistic lighting to make it inviting. Native ferns were selected to withstand the deep shade. The wide, linear seat-wall was well proportioned for the work habits of young tech workers. On any given day you can find co-workers having a meeting with their laptops, meditating or taking a mid-day nap.

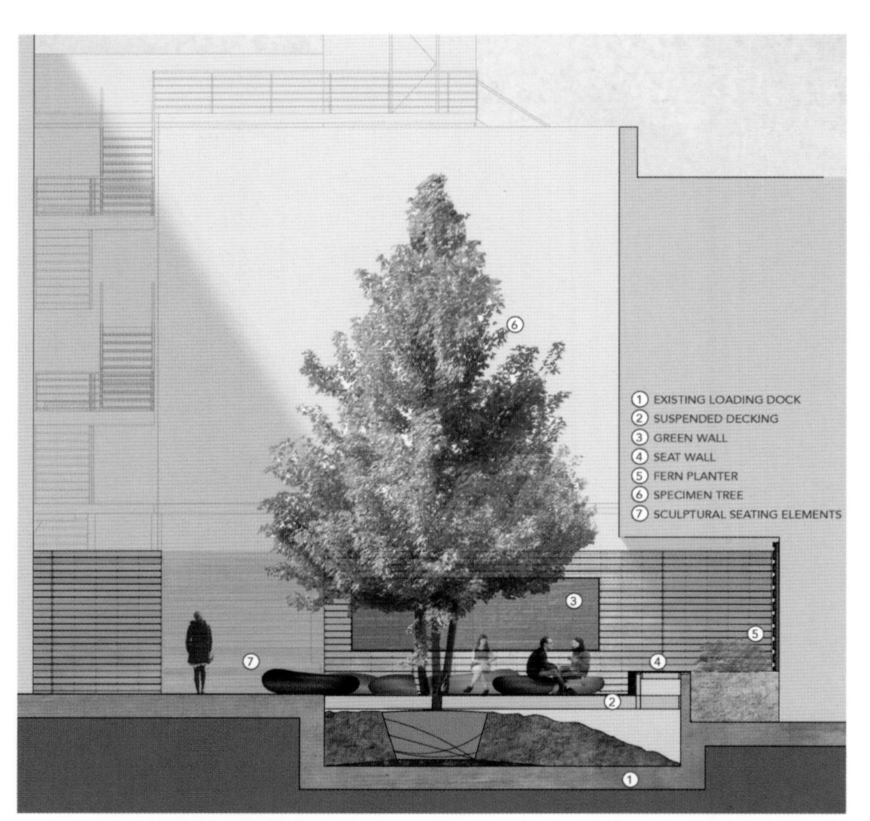

8TH STREET
AIRBNB HEADQUARTERS LOBBY
JEWELRY CENTER
JEWELRY CENTER LOBBY
BRANNAN ST

设计公司 Meyer + Silberberg、Gensler 及 SKS 在旧金山 SOMA 街区重建了空中食宿办公总部的后花园。该设计创建了一个灵活的社会空间，从而创造出一种新的工作方式和合作方式。

本项目的景观设计构想是针对旧金山技术工作者的工作习惯，强调景观的重要作用，并且为快速扩张的城市建设提供绿色空间。原来装卸东西的区域被改造成升高的广场，连接珠宝中心和新空中食宿办公总部。空间不美观之处被彻底改造，打造了一个令人惊叹的舒适景观。

庭院中央是三棵枫树。庭院内设置了木制铺装和桌椅，黑色圆形的石头围绕着三棵枫树，使庭院格外显眼，为人们提供了休闲场所和享用午餐的地方。其他设计元素包括多肉植物墙、大理石铺地走廊、魔法灯饰、木质墙壁。庭院的设计提供了多种结构，反映了非正式的灵活的工作文化。

设计中的一个重要挑战是如何减少悬挑的珠宝中心空间中产生的遮蔽和压抑感。解决方案是在木质墙面的表面加上更多设计细节和艺术灯饰，让整体搭配更为和谐。木质墙体下部还种植了蕨类植物。宽阔的座椅墙按照技术工作者的工作习惯设置。每天，你可以在这里找到用笔记本电脑开会的同事、沉思的同事或打盹的同事。

禅宗对称景观

Zen Symmetry

设计公司：Francis Landscapes

地点： Lebanon（黎巴嫩）

面积：1 000m^2

摄影：Frederic Francis

In such a small (1000 m^2) space threatened by sound and air pollution, as well as the harsh conditions of a hot and moist summer season, the challenge was to create a breathing space in
a front garden, for plants as well as human beings. The outcome could only then be a minimal design intervention, binding the building to the city and allowing for a soft and relaxing transition between the controlled environment of the house and the hectic urban life. A harmonious connection between soft scape and hard scape, a sensitive choice of form, repetition of design elements, a comforting linearity and symmetry, as well as a soothing running-water element, were combined to create a minimal urban garden.

The soft scape is clearly chosen to contrast with the hard scape in form and in color. The small shrubbery is cut into spheres, highlighting the cutting linearity of the basalt slabs. The plant material is native, implying its adaptability to the harsh and changing climate conditions and its requiring less maintenance and irrigation compared to other species.

Using lines and a succession of colors, the garden also presents an extension of the architectural element. Like a carpet, it is a horizontal unfolding of the façade's linearity. This is where the concept of Zen comes to play: "zen" is not taken as the formal replica of the Eastern philosophy;
it is rather regenerated through its underlying laws.

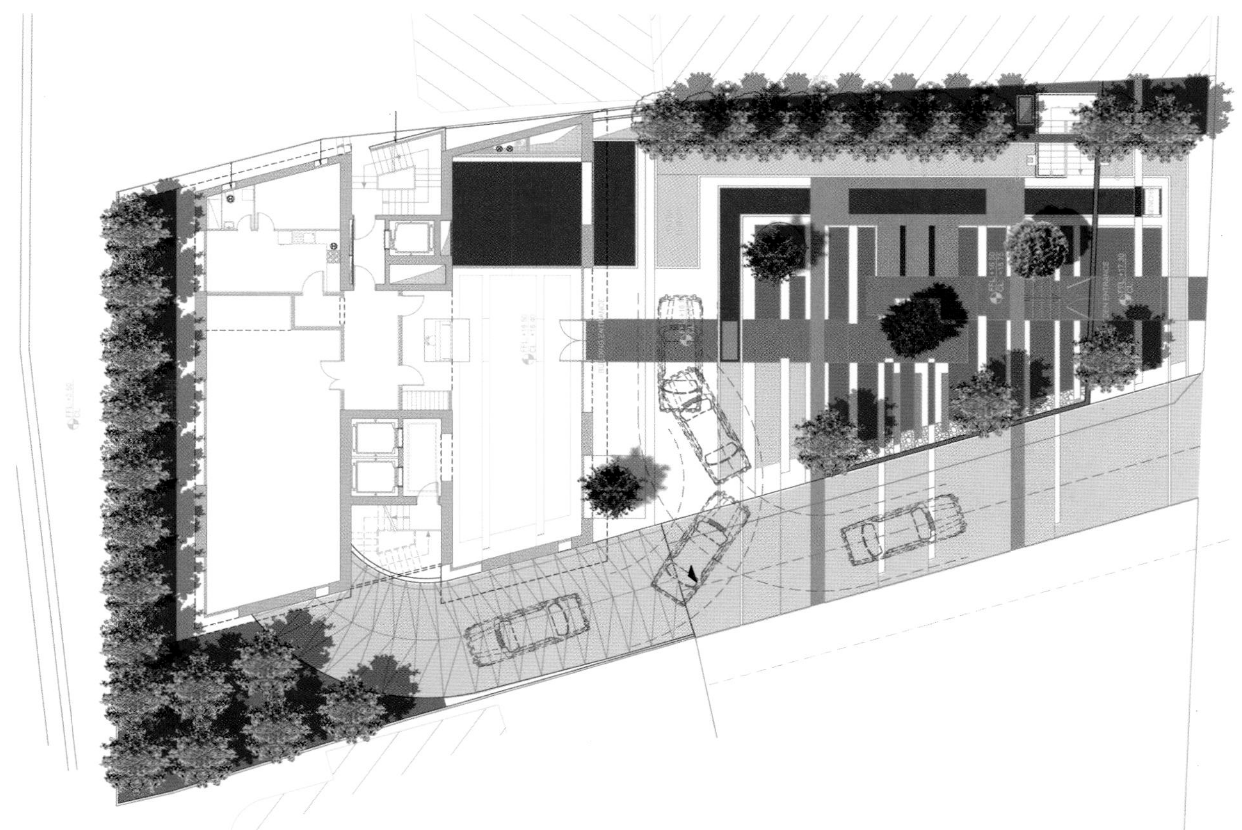

项目位于一块占地面积 1 000m^2 的有噪声和空气污染的地方，这里夏季炎热潮湿，气候条件苛刻。项目设计的目标是为人们建造一个可以呼吸新鲜空气的氧吧。设计将建筑与城市连接，将软质景观和硬质景观完美搭配，并用重复的设计元素、恰当的线形设计及对称设计、流畅的水景元素组成了一个小型的城市花园。

软质景观经过精挑细选，与硬质景观在形式和颜色上形成对比。小的灌木被修剪成球状，突出硬质景观的线形造型。植物使用本地植物，因为它们更适应这里严苛和多变的环境，而且相比其他物种，本地植物需要较少的维护和灌溉。

花园使用了线形造型和丰富的颜色。花园的样子像一个地毯，或是一幅横向展开的绿色的平面图案。这正是本次设计想表达的："禅意不仅是要体现东方哲学，还可以展示事物内在的规律。"

两个空间之间

Between Two Worlds

设计公司： Francis Landscapes

地点： Lebanon（黎巴嫩）

面积： 1 500m^2

摄影： Fares Jammal

Before embarking on this project there was one simple precondition: to make the lobby an integral part of the garden. As such, the relationship between the garden and the building were to be central to the entire project. The entrance is a glass box that extends from the building and reaches into the Japanese inspired green minimalism of its garden. Crowning this lobby is a roof garden to guarantee that no green space is lost, and allowing it to merge with the garden in a three dimensional way.

Creating maze-like pathways and avoiding a straight and overly efficient walk permits this relatively small garden to unfold and whisper its secrets as wanderers stroll through. It also defines and reinforces the relationship between the wanderer and the garden, and between the garden and the building, reinforcing the relationship between two worlds. Graphic and manicured, the plants resemble that of architectural features in their precision, geometry and color.

项目的一个准备工作是将休息室纳为花园的一部分。如此一来，花园和建筑的关系成为项目的焦点。入口处设置了水池。休息室的顶部是一个屋顶花园，为人们提供了更多的绿色空间，整个花园形成一个绿色的三维式的结构。

迷宫式的道路设计避免人们直接抵达目的地，这样的设计使得小巧的花园成为人们交流或分享秘密的场所。此外，设计加强了散步者和花园的联系，同时加强了花园和建筑这两个空间之间的联系。按照几何图形修剪的植物在位置、形状、颜色上体现出花园的景观特色。

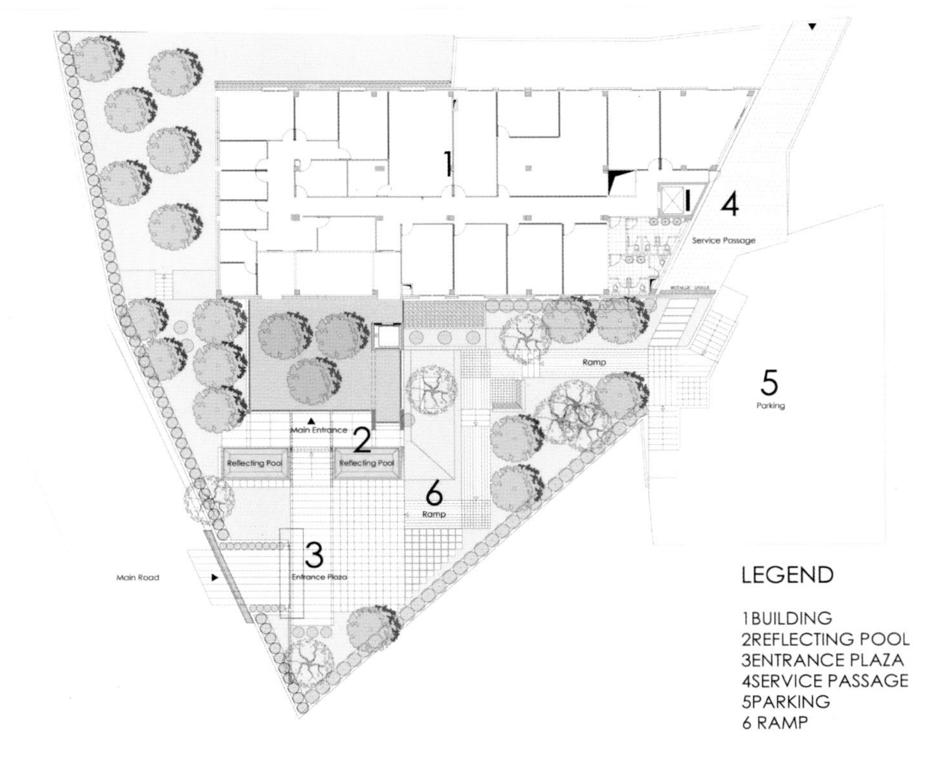

线形透明景观

Linear Transparency

设计公司： Francis Landscapes

地点： Lebanon（黎巴嫩）

面积： 400 m^2

摄影： Frederic Francis

Linear, transparent, detached, and lush a few kilometers out of the city of Beirut, in a green patch in the sky sits the linear garden of a prominent design firm. The stimulating detail in this designed space is the liaison amongst the basic natural elements including air, water, earth, and metal and their linkage between the internal offices and external green spaces.

Enclosed by glass walls at one extremity and a wooden planter of metrosideros hedge the opposite edge, the visitor is engulfed and detached in this 400m^2 roof garden in which one is surrounded by specimen olive trees on either extremity, three carefully punctuated citrus trees, as well as rectangular and circular planters home to various native species fluctuating in color and texture, as well as the numerous species of cactus that decorate the terrace and give it a rather edgy character. In addition to the harmony of soft-scape elements exists a hard-scape interplay between the concrete, the black/white cobble stones and the wooden deck at the commencement of the terrace, plays well in accord with the contrasting shapes of the plants flourishing in this green pocket.

Four seating areas varying in character lie along this outdoor corridor. At the two extremities sit decorative water features imaging the sky and the perforated metallic shade structure that hovers above. The interchange of shadows between the inside and outside allows for the experience of a bold linear transparent space, dynamic and interactive both visually and spatially.

From the inviting bench, the well scattered pebbles, prominent trees, the cactus

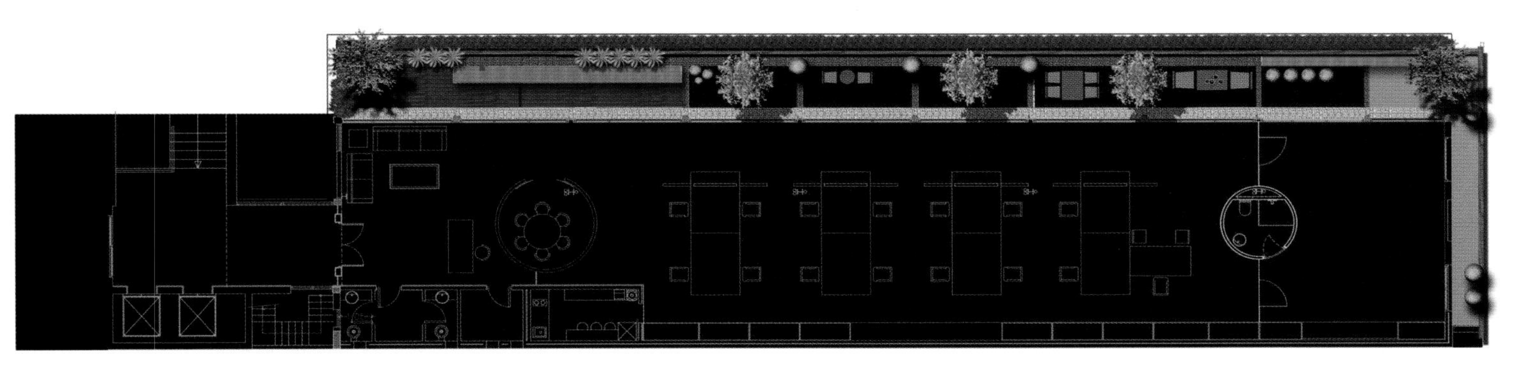

corner, and lush vegetation accented by the omnipresence of water and natural light; all elements interconnect with the symmetrical architecture of the loft which will without a doubt leave an impression on all those who visit it.

黎巴嫩贝鲁特城外 1km 左右，有一处线形的、透明的、独立的、草木丛生的景观。景观矗立于绿色的草地中，由卓越的设计公司设计完成。项目中吸引人的细节部分是自然元素——空气、水、土地、金属等，与内部办公空间和外部绿色草坪之间完美的搭配。

建筑外墙的一部分使用了玻璃材料，玻璃外墙对面的墙边种植了一排乔木作树篱。游客们站在这座占地 400m^2 的屋顶花园中能欣赏到郁郁葱葱的植物景观，3 株精心栽植的柑橘树和矩形、圆形的当地植物盆栽完美搭配。品种多样的仙人掌装饰着平台，使平台变得十分独特。除了柔和的植物元素外，一些硬质元素，如混凝土、黑白鹅卵石、木质平台等，恰当地装饰着整个绿色的环境。

室外的走廊设计了 4 个座位。建筑两个角落处的水景设施倒映出天空的影像。大胆的线形、透明的空间设计呈现了倒影的变化，创造了丰富的视觉体验。

广场椅、散落的鹅卵石、高大的树木、品种多样的仙人掌、茂盛的植被、水和自然光等元素与建筑结构相融合，这将毫无疑问地给参观的人们留下一个深刻的印象。

100+
LANDSCAPE
DESIGN

住宅
RESIDENCE

金斯住宅

Kings House

设计公司： THE PURPLE INK STUDIO

地点： India（印度）

面积： 0.74hm^2

摄影： SHAMANTH PATIL

Two blocks were planned to house one apartment on each side per floor that would emerge from sunken gardens and blend into the peripheral greens amidst the site. Each block was articulated using the existing vegetation as a stencil and building was thus carved out. To compensate on the loss of lower vegetation from the site during the construction, every floor plate extends into greens and balconies generating great diversity within the site context.

An integrated design approach was followed to evaluate and maximize the energy reductions of the building. To optimize the cooling effect, the building mass and window openings were shaped and sized to best capture the breezes based on Computer Generated Simulations. The Vertical Shading devices are combined with Horizontals to cut off harsh rays of the sun. This functions as a rain protector and also multiplies as a Visual Barrier Sideways.

While the concept of the building focuses on green factors of design and use of sustainable materials, the aesthetic character of the building is far from being compromised. The ecological elements are conscientiously woven together with the luxury requirements of the project that conclusively expresses a contemporary response which further establishes a contextual relationship and giving each residence the highest degree of originality.

公寓的两侧分别有一个街区。公寓与下沉式花园相连接，附近芳草茵茵，绿树葱茏。每个街区内都分布着绿色的植被。为了补充建设而损失的较低矮的植物景观，建筑每层地面延伸出一段距离，形成花园阳台景观。

设计师使用集成式设计方案，将建筑的能源减排最大化。项目充分利用冷效应，并借助电脑生成模拟系统，可将建筑总面积控制在最佳大小，窗户开口控制在最佳形状，让更多自然风进入到室内。垂直及水平阴影装置可以抵挡住强烈的阳光照射。此外，它也可以抵挡雨水。

建筑理念集中在绿色设计和可持续材料的使用上，同时保证不破坏建筑的美观。生态元素穿插在设计中，表达了当代理想城市的理念，给居民们创造一个高度绿色的环境。

Amenity Block +
Swimming Pool above
Central Atrium with
Mounds + Waterbodies
Peripheral buffer planting
Green Courts +
Barbeque Areas
Fruiting trees
Wing B Lobby
Peripheral buffer planting
Concentric Court
Peripheral walkway
Wing A Lobby
Pedistrian
ENTRY
Vehicular ENTRY
to Basement

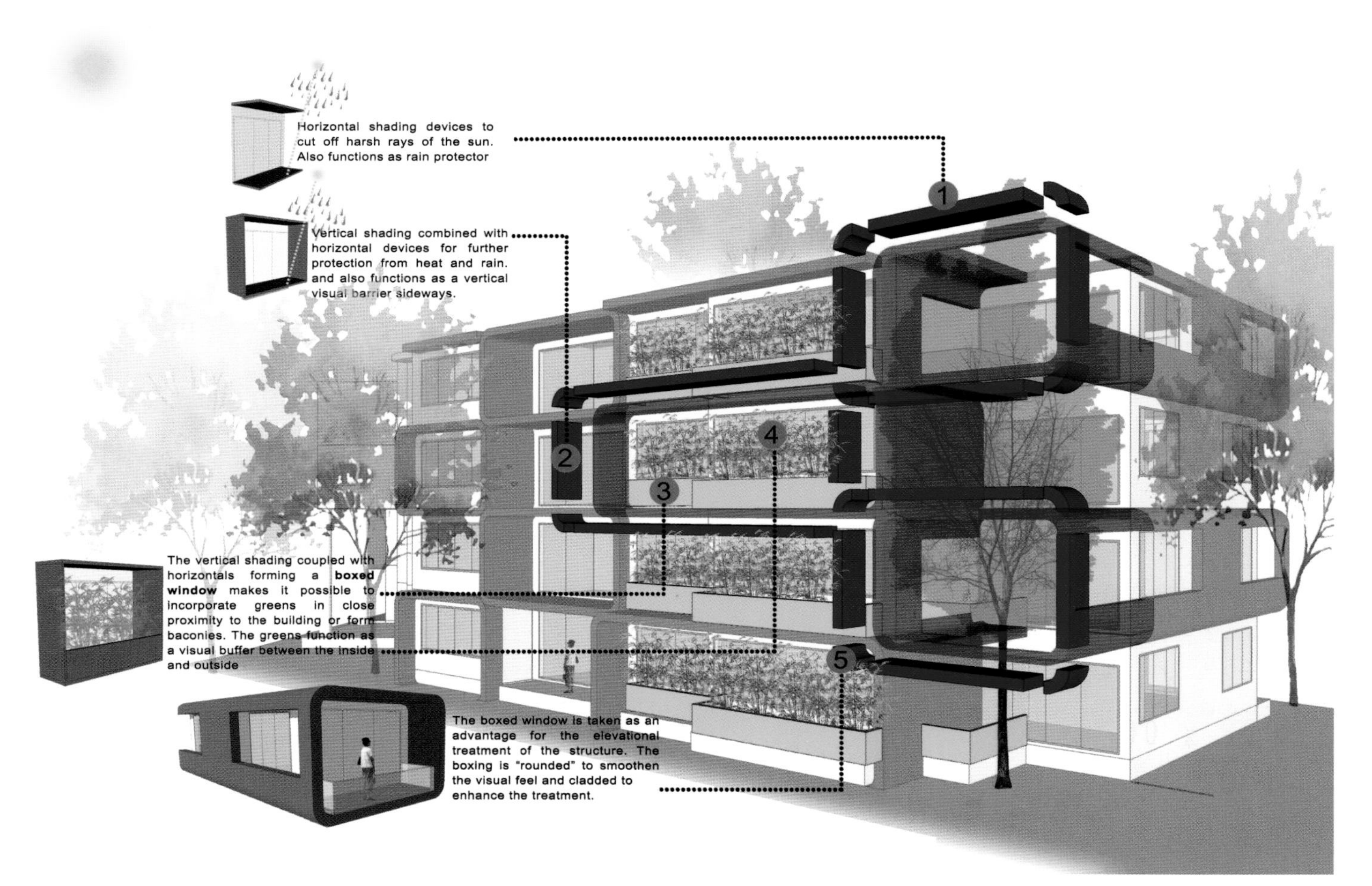

Baan San Kraam 住宅景观

Baan San Kraam

设计公司： Sanitas Studio

设计团队： Sanitas Pradittasnee(Design Director), Ronarong Chompoopan, Supavadee Nimawan, Rachaniporn Tiempayotorn, Amisa Raksiam, Vongvaritt Siwatwarasuk

地点： Thailand（泰国）

面积： 23 109.96 m^2

摄影： Wison Tungthunya

Starting with a nautical nostalgia theme, Sanitas envisaged the land as an abstract form of the ocean. It would contain different elements of islands and seascape such as jungle, villages and a floating house.

Seven clusters of buildings were then created with their own unique seascape character.

The landscape design is a simple interpretation of the wave typology. Having studied its form in detail, Sanitas has developed it into three dimensional landscape form.This includes wave seating, stepping stones, rock day beds and a tree house, while water is the landscape's key element and connects all zones together.

Residents can experience the different character of each area from their arrival at Lobby Beach and then across the water to the Jungle which is surrounded by secluded villages. Local coastal plants enhance the natural beauty of the development and its stunning oceanside concept, while existing trees have been preserved to provide welcome shaded area.

该项目是以船员的思乡之情为主题而建造的。Sanitas 设计团队将该场地当成大海的缩影进行设计。该景观设计中包含了各式各样的岛屿景观元素和海景元素，比如茂密的丛林、古朴的村庄、水面上的漂浮住宅。

七个楼群陆续崛起，造型特点各异，构成了一幅独特的海边景观。

景观设计再现了海滨的波浪的形态。经过细致的分析测算，Sanitas 设计

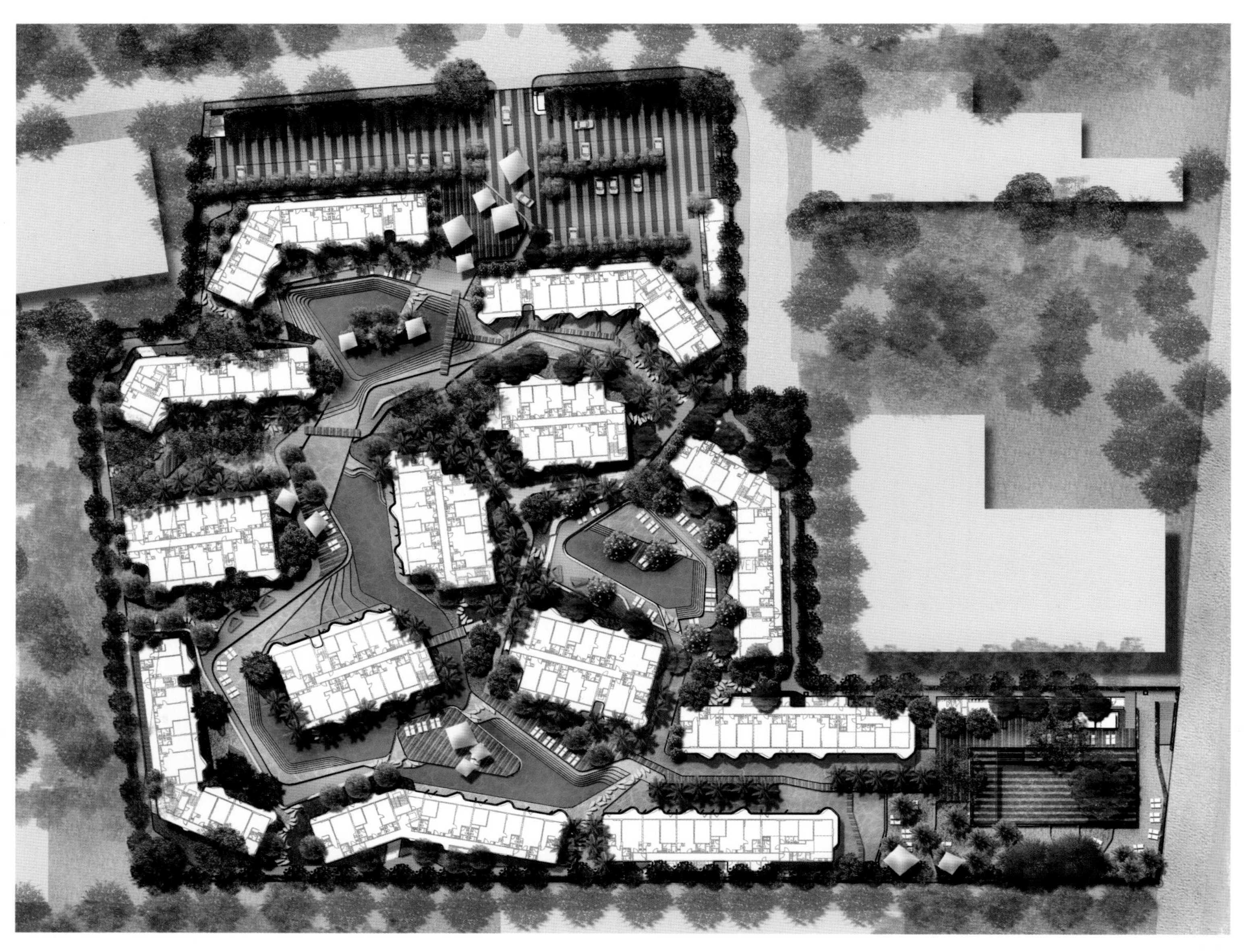

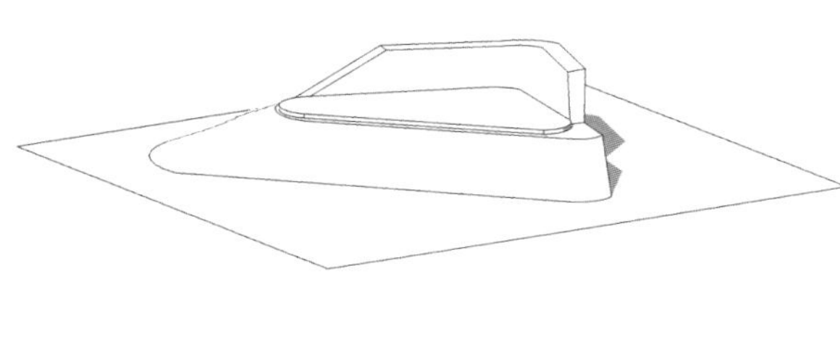

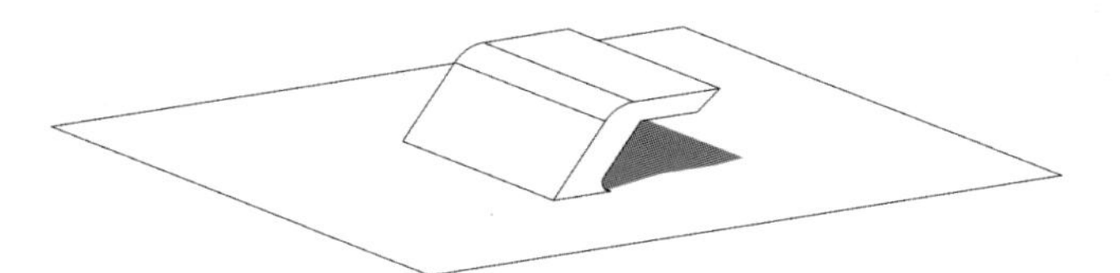

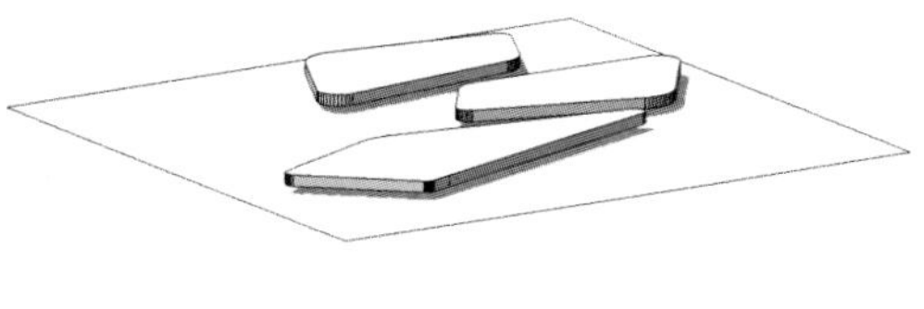

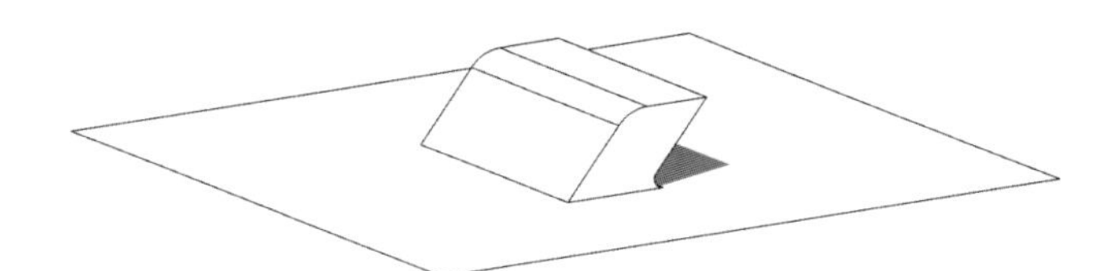

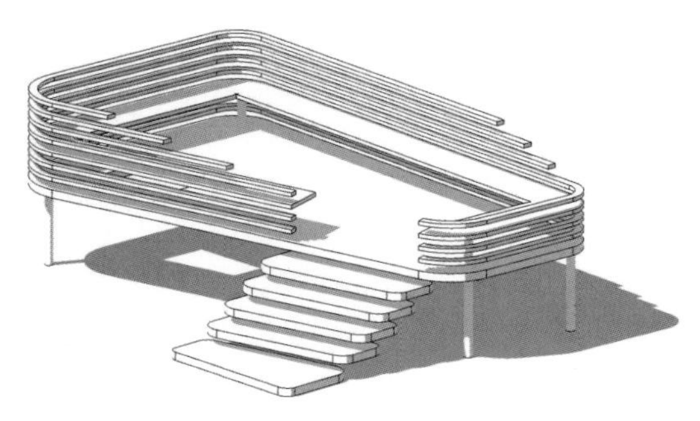

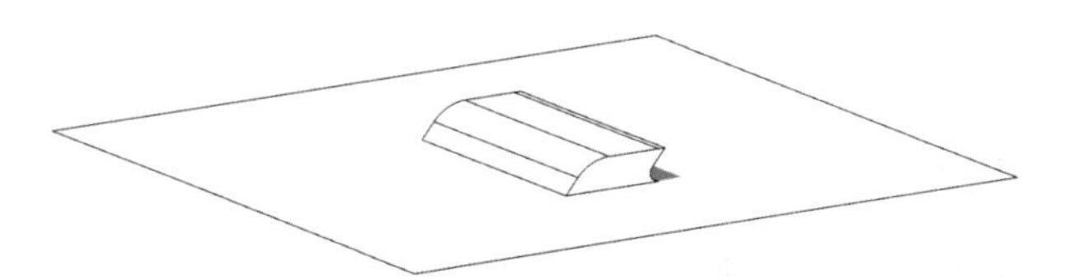

团队力求使该景观呈现出三维立体效果，为达到这个目标，他们设计了波浪状的座位区、石板铺成的路面、白色简约的躺椅及一个树屋。在该景观中，水景是关键性的装饰元素，它将场地中的各个区域紧密地联系在一起。

游客们只要踏上这片海滩，便会被这里每个地段独特的美景所吸引，他们可以穿过美丽的水景到达茂密的丛林，丛林四周是隐蔽的小村庄。此外，当地的沿岸植物为该项目整体景观增添了一份自然魅力，并成为优美的海岸景观的一部分。值得一提的是，海滩上本来就种有大树，它们为人们提供了乘凉的地方，吸引了更多的人来到这里参观。

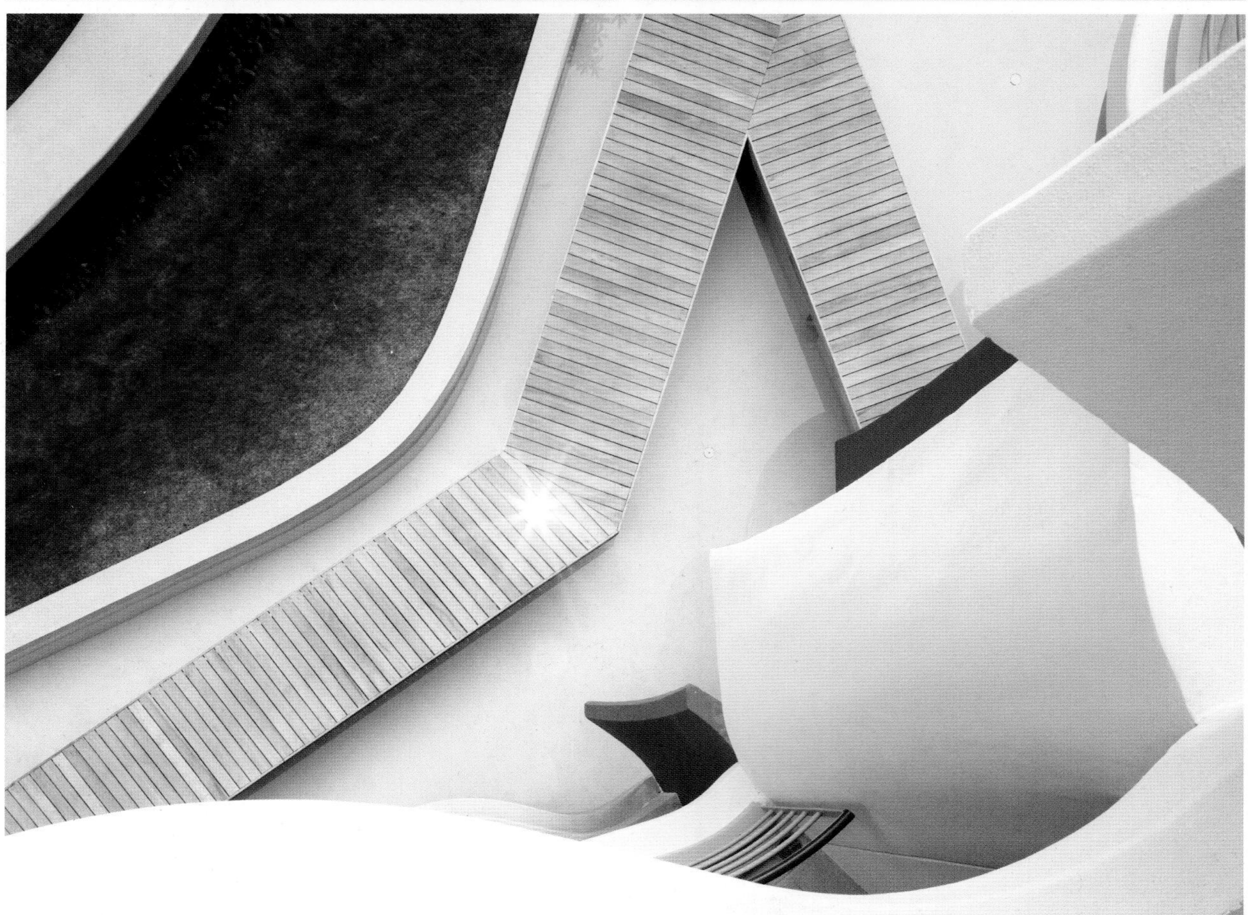

南洋理工大学学生宿舍

Ntu Crescent Hall and Pioneer Hall

设计公司： RSP Architects Planners & Engineers (Pte) Ltd in collaboration with Toyo Ito & Associates, Architects (TIAA)

地点： Singapore（新加坡）

面积： 34 967.98 m²

效果图表现： Kuramochi+Oguma, Toyo Ito & Associates

摄影： RSP Architects Planners & Engineers

An Extended "Forest" within the University's Garden Campus

Within the well-established green network at NTU's Garden Campus, we proposed eight tree-like blocks each with a three-directional planar formation that resembles branches stemming from a tree trunk, challenging the norm of a typical slab-block typology for student dormitories. Circumventing the size constraints of the site, the tree blocks are laid out in a way that maximise their interaction with greenery, wind and water, offering a comfortable living and learning experience akin to sitting under a large shade tree.

A New Form of Integrated Community

Residential units housing 1250 students start from the 3rd storey onwards, with each block carefully positioned to preserve occupant privacy and offer unique views despite the density of the premise. Shared TV lounges and small pockets of gathering spaces are built on every floor of the centrally located "trunk" for easy accessibility. Larger common spaces such as multipurpose halls and canteens are located on or close to ground level and woven into the verdant surroundings.

Environmentally Sensitive Design Features

The existing site has a stark level difference with an open, swampy drain that cuts across the valley of the site. The design attempts to respect the existing terrain and transform the existing drain into an extended water body that comprises a sedimentation basin, wetlands, rain gardens, a

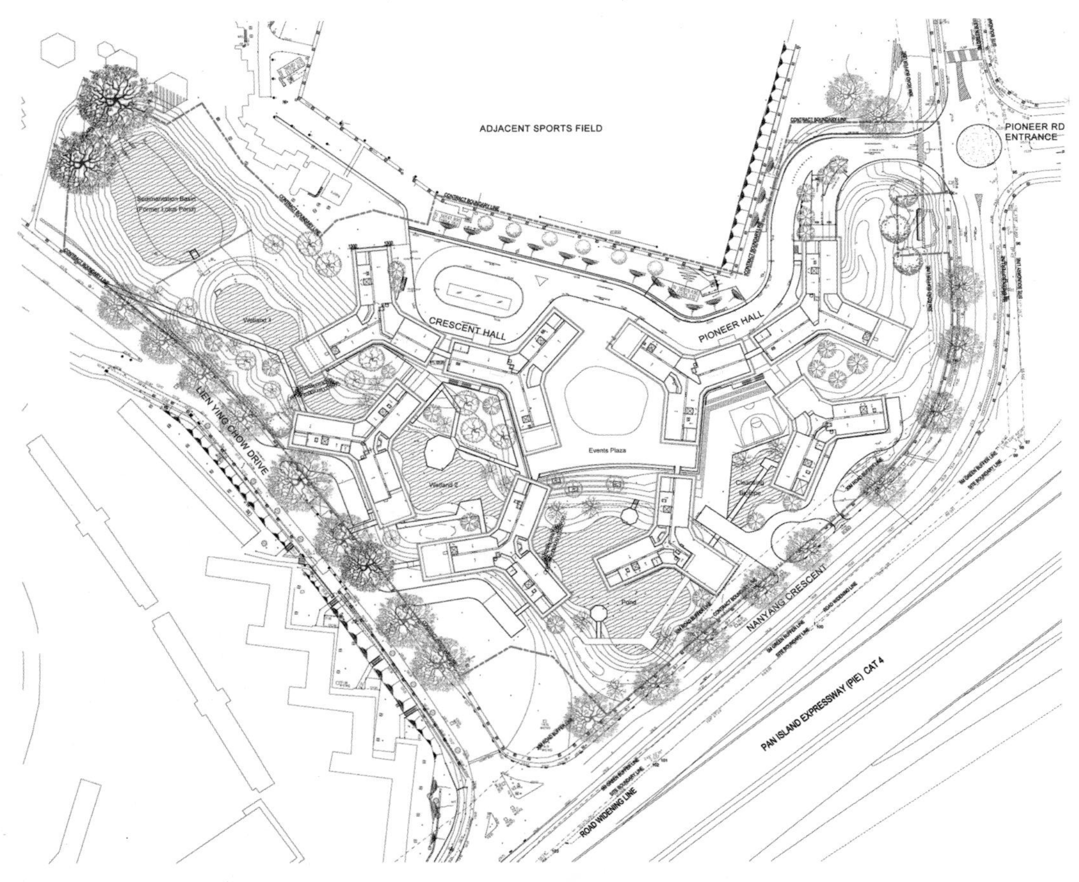
ADJACENT SPORTS FIELD
PIONEER RD ENTRANCE
CRESCENT HALL
PIONEER HALL
Events Plaza
LIEN YING CHOW DRIVE
NANYANG CRESCENT
PAN ISLAND EXPRESSWAY (PIE) CAT 4
ROAD WIDENING LINE

17B
CRESCENT HALL

cleansing biotope and a retention pond. These water bodies are not only visually pleasing but also act as a holistic treatment train that cleanses storm water before it enters the public drains, further enhancing the national water security and catchment principles. The integration of naturalised water bodies and living quarters create a convivial and sustainable environment where flora and fauna can thrive. While this feature serves as a campus educational tool on natural habitats, it is a potential prototype for future private/public residential developments locally and regionally.

大学校园里的“热带雨林”

南洋理工大学的新花园校园中有 8 个类似于树状结构的新建筑。建筑外形十分奇特，是对传统的学生宿舍建筑的挑战。树形建筑绿色环保，同时拥有最大化通风和通水性能，学生可以获得在树荫下学习和生活的体验。

新形态的联合社区

新宿舍从第三层起可以容纳 1250 名学生，如此高的密度并不影响每栋建筑中留给学生的个人空间，同时提供了绝佳的景色观赏台。建筑每层都设有公共的电视房和小型聚会房，它们处在靠近建筑中央的位置，方便出入。大型的多功能厅和餐厅位置较低，靠近一层，掩映在周边郁郁葱葱的环境中。

设计特点——环境敏感型

这里曾经使用的是开放的、潮湿的排水系统，与新系统有很大差别。新的设计在旧的排水系统的基础上，加入了自然生态的沉淀池、湿地、雨水花园、清洁区、蓄水池。这套生态排水系统不仅外观与周围景观融为一体，而且能够在雨水进入公共排水系统之前过被收集，提高水的回收效率。生态水体和生活区的结合创造了可持续的植物生长环境，成为优秀的生态教科案例。在未来，这将是个有潜力的发展模式，可以用在私人及公共住宅区。

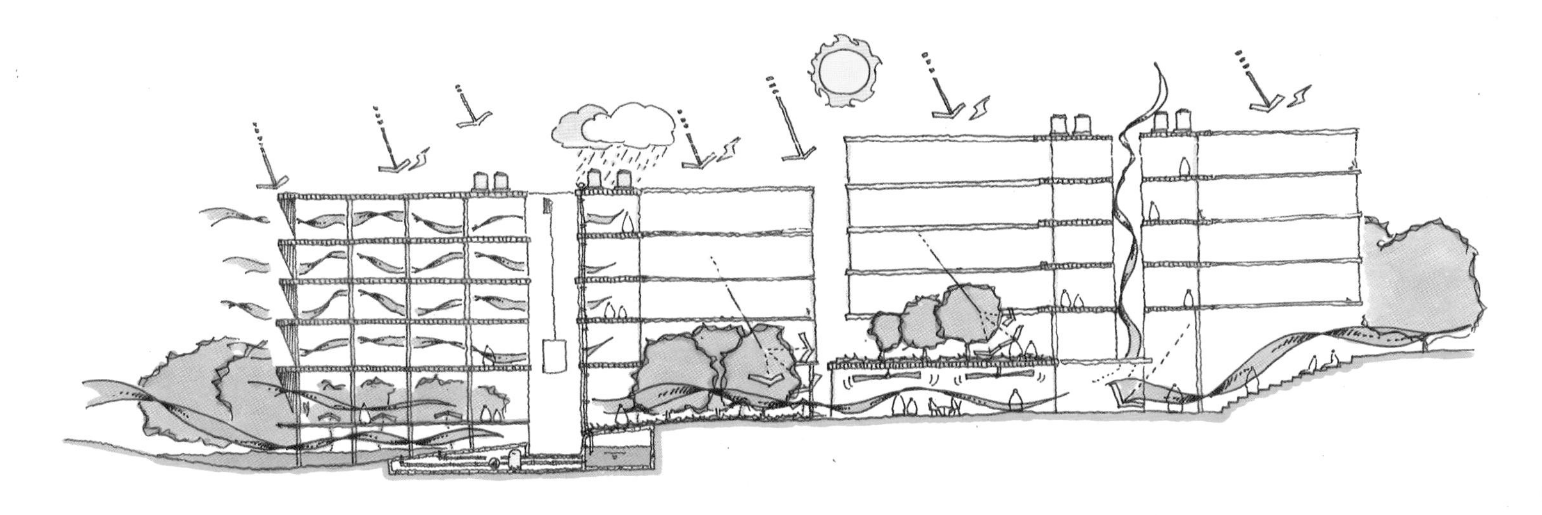

心怡苑住宅

NV Residences

设计公司：Salad Dressing

地点：Singapore（新加坡）

面积：27 000 m^2

摄影：Mr Ng Sze Oun

This is a high-end residential project, the landscape design expresses the theme of tropical rain forest. The designer finally creates a residential space combining leisure and recreation.

Designer chooses big arbors to make a tree array which combines together with the building's style and the floor pavement. Under the tree array, there is a row of manicured hedge. Those compose typical regulated design. The design of leisure and recreation area uses natural design style. Round and bulging or sunken green lawns are dotted on a large plastic cement ground. 40cm-long green retaining wall segments the space and the wall's below is been opened to add interest to the space. There are lots of various play equipment in here and it is the best recreation place for children.

这是新加坡的较为高端的住宅项目，景观设计体现了热带雨林的主题。设计师最终打造了一个集休闲娱乐为一体的住宅空间。

设计师选择用高大的乔木做树阵，这与建筑的风格和地面的铺装融为一体。树阵的下方是修剪整齐的绿篱。这些组成了典型的规则式设计。休闲娱乐区的设计则采用了自然式的设计，圆形的或凸出或凹陷的绿色的草坪点缀在大面积的塑胶地面上。40cm 左右的绿色的挡土墙又将空间细分化，挡土墙的下部还做了开口处理，为空间增添了趣味性。这里放置了多种多样的游乐器械，是儿童们玩耍的最佳场地。

Minton 住宅

The Minton

The Minton is completed in the year 2014. It comprises ten 15-storey and eight 17-storey apartment blocks. The designer makes full use of the site topography. Landscape design gives people totally different experiences. The blocks are connected by bridges and sky terraces. The Minton not only provides basic amenities but also much wider facilities' service design in a delicate space. Luxuriant vegetation creates a cozy and private living environment which seems like a peaceful urban house.The design reflects the increasingly sophisticated demands of the modern people and creates a whole new living style.

设计公司：Salad Dressing

地点：Singapore（新加坡）

面积：47 000 m^2

摄影：Mr Ng Sze Oun

Minton 住宅于 2014 年建成。由 10 栋 15 层楼和 8 栋 17 层楼的公寓大楼组成，设计充分利用了场地地形，景观设计给住户提供了完全不同的居住体验，木质桥梁连接起了各栋大楼。Minton 提供的不仅仅是基本的设施，还在一个经过精心设计的空间中提供了更为广泛的设施服务设计。郁郁葱葱的植物创造了一个舒适而宜居的私人居住空间，是城市中的一所静居。设计反映了现代人不断增长的复杂需求，并创造了一种全新的生活方式。

TREEHOUSE

Domain 21 公寓

Domain 21

设计公司：Salad Dressing

地点：Singapore（新加坡）

面积：5 357 m^2

摄影：Mr Chang Huaiyan

The project is a design for part of residential area. Designer designs different functional spaces in the limited room. There are gallery frame space surrounded by greenery, pool space available for play, and road space with stone and dotted bar grass land. All the spaces have their own characters. The match of the community landscape on the whole are perfect. Designer creates peaceful and cozy space, giving the residents more space for communication. People can do exercise or have a cup of tea or have a chat or enjoy a good life here.

本项目是对居住区的一部分空间进行设计。设计师在有限的空间内设计出不同功能的空间。这里有被绿色乔木围绕的廊架空间，有戏水的泳池空间，还有铺设了石板、中间点缀条形草地的道路空间。所有的空间都各有特点，又从整体上与社区的景观完美搭配。设计师创造出幽静、舒适的空间，给社区的居民们提供了更多互相交流的空间。人们可以在这里健身，或是喝茶、聊天，享受惬意的生活。

康科德住宅区

Concord Residences

设计公司：Landscape Architecture Bureau

地点：USA（美国）

面积：0.61 hm^2

摄影：Prakash Patel, Maxwell McKenzie

The courtyard at the Concord Residences is an intensive green roof over a below-grade parking structure. Landscape Architecture Bureau worked closely with the project's developer and architect to design a courtyard that was suitable for multiple uses but that was, at the same time, simple and serene. The courtyard accommodates social gatherings while, at the same time, due to its straightforward geometry and the simple palette of materials, it can also serve as a quiet, calm place to read or have a cup of coffee.

The courtyard is structured as a series of rectangles; the water feature, lawn panel, seating areas, and tree plantings are all in this courtyard. The design also incorporates a simple, passive strategy to re-use all stormwater for irrigation. Rain that falls on the lawn, on the plaza pavement, and that runs down the facade of the building is captured in trench drains that supply a system of perforated pipes under the pavement. The water is distributed through this system into vaults under the courtyard filled with enriched planting soil.

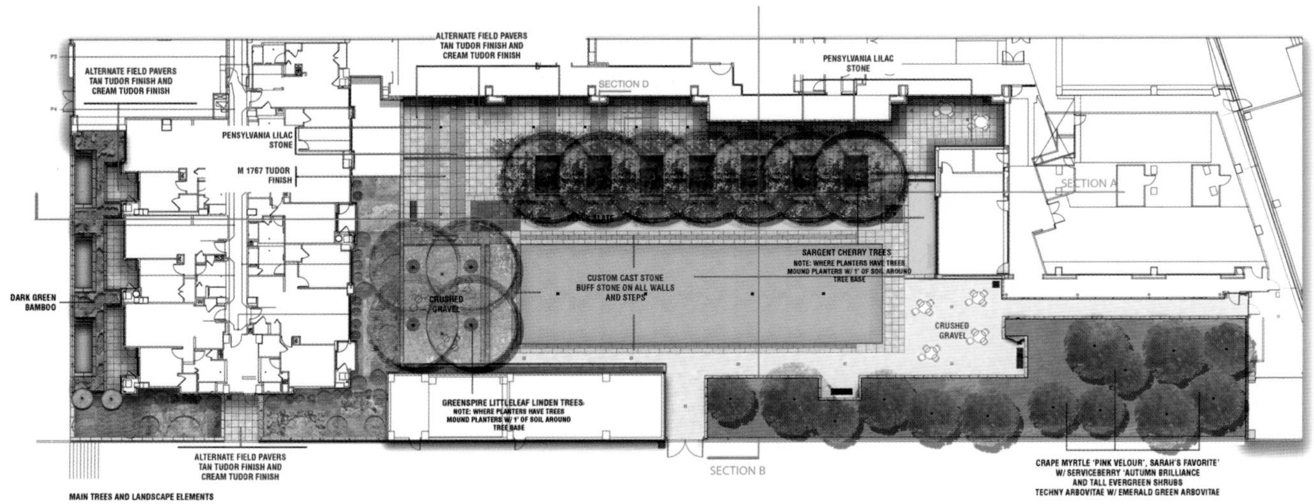

MAIN TREES AND LANDSCAPE ELEMENTS

康科德住宅区空间更像是一处植被繁茂的绿色屋顶花园，这里拥有一个地下停车场。景观设计师和项目的开发商通力协作，设计了这里的景观，虽然有一些单调，但是用途广泛，为社团集会提供了场地。另外，景观平面采用直线几何形的形式布局，用材简单，使这里成为可以安静地读书和享受咖啡的地方。

住宅区内设置了人工水景、草坪、乘凉座椅及成片的绿树。设计采用简单的雨水收集的方法，并将雨水重新用于灌溉。落在草坪上、广场石阶上、建筑外墙上的雨水进入排水沟，通过多孔的排水管渗入地面，从而起到灌溉的作用。

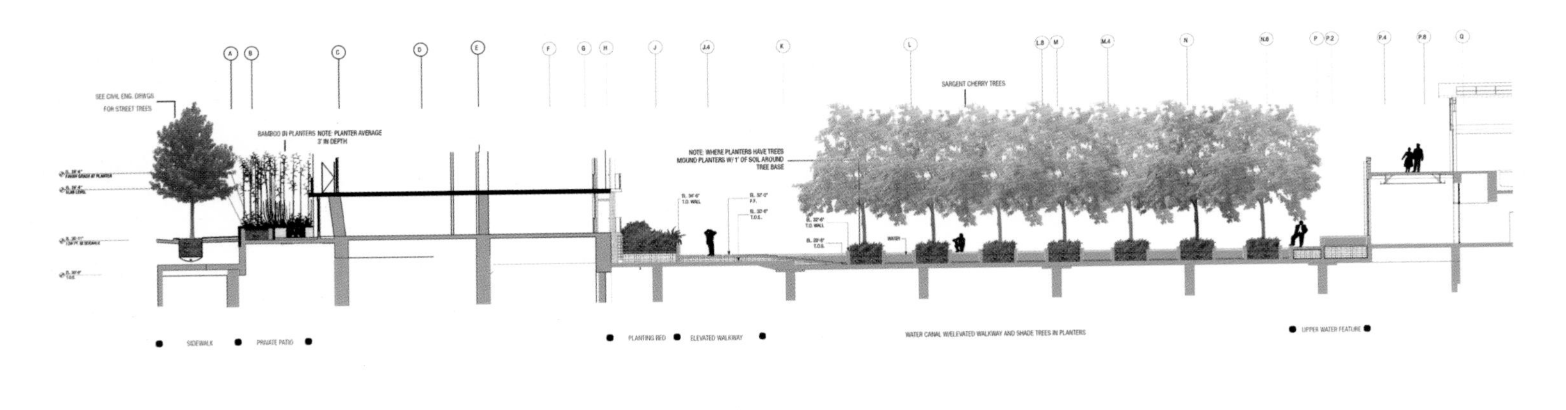
SEE CIVIL ENG. DRWGS
FOR STREET TREES
BAMBOO IN PLANTERS NOTE: PLANTER AVERAGE
3' IN DEPTH
NOTE: WHERE PLANTERS HAVE TREES
MOUND PLANTERS W/ 1' OF SOIL AROUND
TREE BASE
SARGENT CHERRY TREES
SIDEWALK
PRIVATE PATIO
PLANTING BED
ELEVATED WALKWAY
WATER CANAL W/ELEVATED WALKWAY AND SHADE TREES IN PLANTERS
UPPER WATER FEATURE

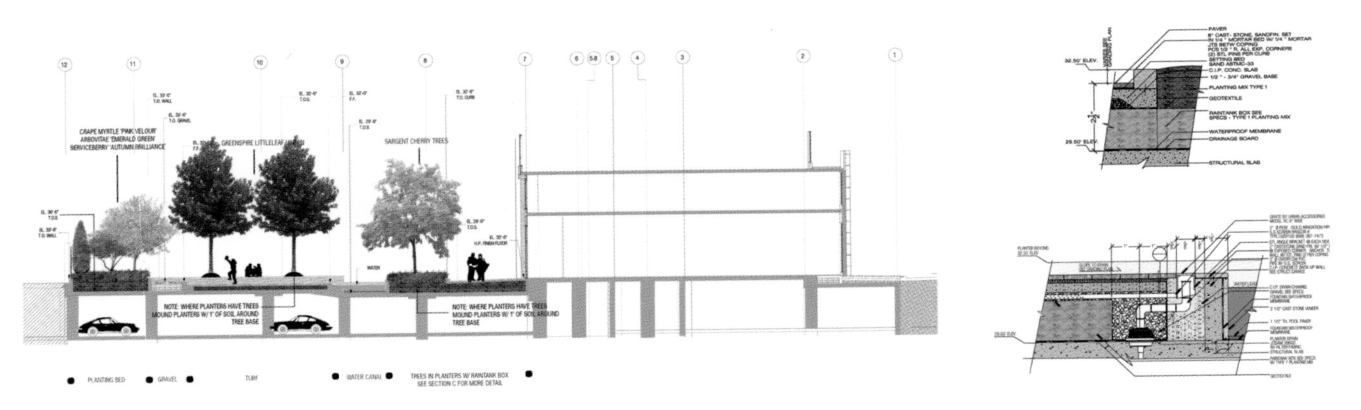
CRAPE MYRTLE 'PINK VELOUR'
ARBOVITAE 'EMERALD GREEN'
SERVICEBERRY 'AUTUMN BRILLIANCE'
GREENSPIRE LITTLELEAF LINDEN
SARGENT CHERRY TREES
WATER
NOTE: WHERE PLANTERS HAVE TREES
MOUND PLANTERS W/ 1' OF SOIL AROUND
TREE BASE
NOTE: WHERE PLANTERS HAVE TREES
MOUND PLANTERS W/ 1' OF SOIL AROUND
TREE BASE
PLANTING BED
GRAVEL
TURF
WATER CANAL
TREES IN PLANTERS W/ RAINTANK BOX
SEE SECTION C FOR MORE DETAIL
32.50' ELEV.
29.50' ELEV.
PAVER
8" CAST- STONE, SANDFIN. SET
IN 1/4 " MORTAR BED W/ 1/4 " MORTAR
JTS BETW COPING
PCS 1/2 " R. ALL EXP. CORNERS
(2) STL PINS PER CURB
SETTING BED
SAND ASTMC-33
C.I.P. CONC. SLAB
1/2 " - 3/4" GRAVEL BASE
PLANTING MIX TYPE 1
GEOTEXTILE
RAINTANK BOX SEE
SPECS - TYPE 1 PLANTING MIX
WATERPROOF MEMBRANE
DRAINAGE BOARD
STRUCTURAL SLAB
2 1/2" CAST STONE VENEER
1 1/2" TH. POOL PAVER
STRUCTURAL SLAB

Kemerlife XXI 住宅区

Kemerlife XXI

设计公司： DS Architecture – Landscape

设计师： Deniz Aslan

地点： Turkey（土耳其）

面积： 4 hm^2

摄影： Gürkan Akay

Kemerlife XXI is a multi-housing residential project which settles 4hm^2. Within the architectural project's style and foresights; in its own environment, project has achieved being the least heat-dissipating and without any loss, rainwater utilizing project. No waste due to landscape is being left, especially when project is located in the area which has slightly environmental issues. With that also a new ecosystem has been created which put forward users' happiness and biorhythm. In the located area, in contrast to the ordinary, environmental issues were dealt in a responsive way. All the terraces were designed as usable gardens and social center was designed as an aqua garden. Not only microclimatic but also both relaxing and entertaining aqua garden help landscape to become an unique contemporary design which does not imitate the nature, but on the other hand, builds a new one out of it. This garden was utilized as house gardens, in-between house pathways, playgrounds and socializing areas. All the materials used in the project were found in the near environment and derived from the processed natural materials in the construction area.

Within the collaboration with client, architectural group, in the whole process, as much as possible minimum construction and operational expenses, easy maintainable and pleasant living environment were main aims to be achieved.

Inner courtyard

A. Swimming pool
B. Sunbathing platform
C. Wooden deck
D. Grass platform
E. Playground
F. Kids' swimming pool
G. Water plants
H. Pebble stone terrace

House gardens

1. Travertine terrace
2. Wooden deck

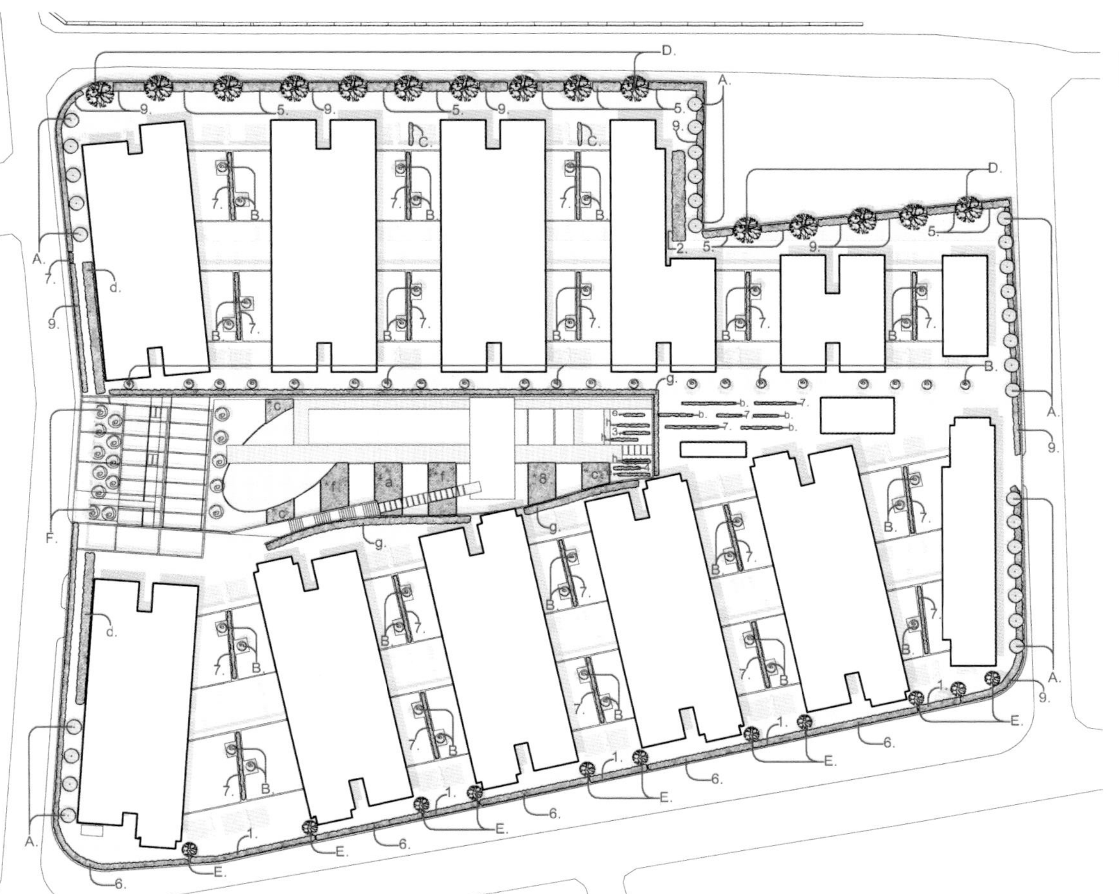

Trees

A. *Acer platonoides "Crimson king"*
B. *Lagerstomia indica*
C. *Laurus nobilis*
D. *Magnolia grandiflora*
E. *Prunus serrulata*
F. *Quercus ilex*

Shrubs

1. *Abelia grandiflora nana*
2. *Bambusa nana*
3. *Calla palustris*
4. *Carex elata*
5. *Escalonia bifida*
6. *Ligustrum texanum*
7. *Photinia fraserii 'Red Robin'*
8. *Rosa rugosa*
9. *Viburnum lucidum*

Perenials

a. *Agapanthus africanus*
b. *Berberis thungbergii Atropurpurea*
c. *Coreopsis sps.*
d. *Cotoneaster salicifolia*
e. *Iris pseudocorus*
f. *Lavandula angustifolia*
g. *Teucrium fruticans*
h. *Typha angustifolia*

kemerlifeXXI
kemerlifeXXI

土耳其 Kemerlife XXI 住宅区景观面积为 4 hm^2，这片小区拥有多栋住宅楼。整个住宅区景观有独特的风格，最大程度地减少了热量的消耗和水分的流失。虽然小区内有一些环境问题，但是这没有影响设计师设计节约型的景观。小区建起来后，考虑到居民的居住满意度和舒适度，设计师在小区中加入了新的生态系统。接下来，设计师着手处理明显的环境问题。每家每户的阳台被设计成多用途花园，社交中心被设计成水景花园。微气候和娱乐水景花园组成了一个特殊的现代设计景观。新设计摒弃了对环境的单纯模仿，在建筑区之外创造了一个新环境。水景花园可以作为社区大花园，与小路、游乐场和社交区有机地结合。场地中的所有景观材料都来自当地。

建设过程由客户、设计师团队通力合作，最大程度减少开支。设计以建造易于维护的和谐人居环境为最终目标。

乌鲁斯·萨瓦住宅

Ulus Savoy Housing

设计公司： DS Architecture – Landscape

设计团队： Deniz Aslan, Özge Akaydın, Selda ipek

地点： Turkey（土耳其）

面积： site 6 hm², landscape 3.5 hm²

摄影： Cemal Emden

Ulus Savoy Housing is a special project, with its fractured structure of parking ceiling which gives a unique character to the site and yet, that forms the base/ shell of the landscape.

Ulus Savoy Housing which is located near to Istanbul Bosphorus region, beside the dynamic structure of the topography, acquires the chance to have the excellent Bosphorus views. It rests on a total area of approximately 6hm² and 3.5hm² of it is designed as open spaces. Ulus Savoy is a multi-housing project with 26 blocks settled on a garage structure which is the substructure of the new landscape topography as well. By means of sharp and hard parts, the amorphous structure of the garage (the shell) is used to embody the landscape. While some parts of the shell are covered with vegetation, other parts are paved with natural stone. On the other hand, this shell works as the backdrop for all of the connection axes and recreational activities. In other word, the constructed topography of the settlement shapes the landscape/garden features of the common spaces. Throughout the project, flat areas serve as the private gardens and the recess areas host the social facilities. Providing a promenade to stroll around the gardens, the partially elevated path plays a great role in the perception of the fractured surfaces of the site.

The steep fragments of the fractured surfaces which are covered by natural stone with particular details. As a result of the seasonal changes, throughout the plantation design, dramatic oppositions create dynamic

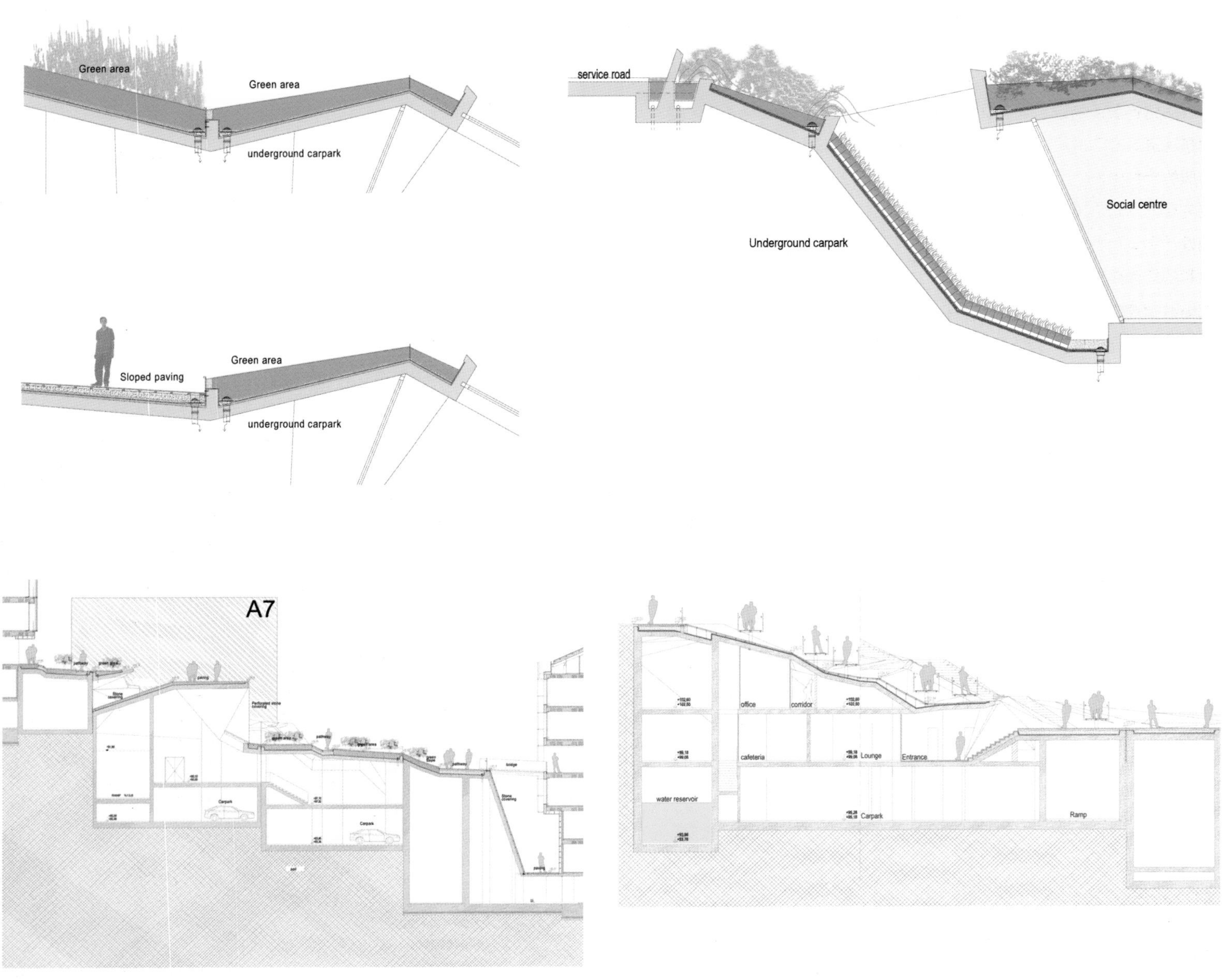

Green area
Green area
underground carpark
service road
Social centre
Underground carpark
Sloped paving
Green area
underground carpark
A7
office
corridor
cafeteria
Lounge
Entrance
water reservoir
Carpark
Ramp

snapshots.

The entrance facade of the settlement contains a permeable linearity within the architectural setting. The metal pots that are added to the metal construction elements (which we use in many of our projects) are designed and placed especially for this project. Starting from the entrance façade, a soft lighting effect is provided and spread throughout the whole landscape.

By the virtue of its unique architectonic and spatial features, this new experimental settlement of Istanbul gains a special place within the city. The theme of the public park which is considered as an extension to the project is an edible garden/landcape. The park is designed with a close approach to the main project idea; the agricultural landscape.

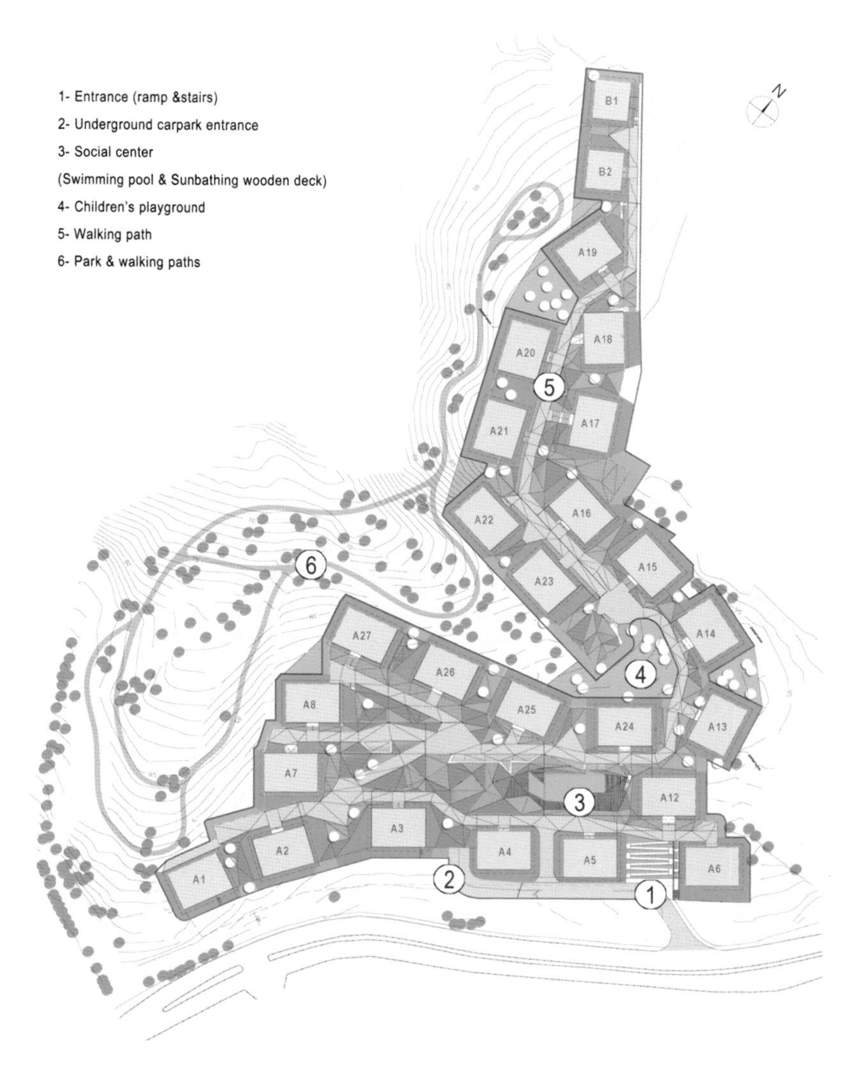

乌鲁斯 · 萨瓦住宅是一个特别的项目，项目中的车库位于断裂的结构上，起伏的地理结构是项目设计的基础和依据。

乌鲁斯 · 萨瓦住宅区位于伊斯坦布尔博斯普鲁斯海峡地区，景色宜人，地形富于动态。占地面积约 6hm^2，其中 3.5hm^2 为开放空间，是一个涵盖 26 个街区的多住宅项目。车库的不规则结构锐利、坚硬，构成了景观的有机组成部分，其表面或被植物覆盖，或铺有天然石材，塑造了公共空间景观的特色。整个项目中，平坦区建设为私人花园，凹陷区则搭建公共设施。一条长廊围绕花园而建。

断裂的结构表面铺设了天然石材。随着季节变化，这些石材表皮与植物景观形成鲜明对比。建筑结构大量运用木材和石材，可谓是对自然本质的全新诠释。

住宅区的入口立面设计考虑了建筑内部采光和通风的需求。金属元素的设计和运用独具匠心。从入口至整个住宅区景观，设计师都采用了柔和的灯光效果。

凭借其独特的建筑和空间特征，这种新的住宅区成为伊斯坦布尔的设计典范。农业景观是该项目的主要设计理念之一。公共花园的主题与项目整体密不可分，设计师在花园中种植了可食用的农作物。

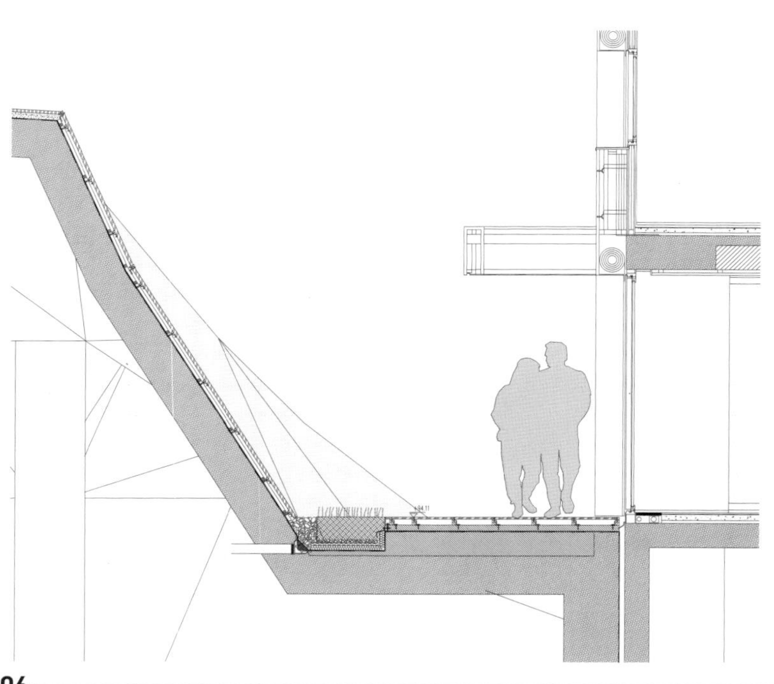

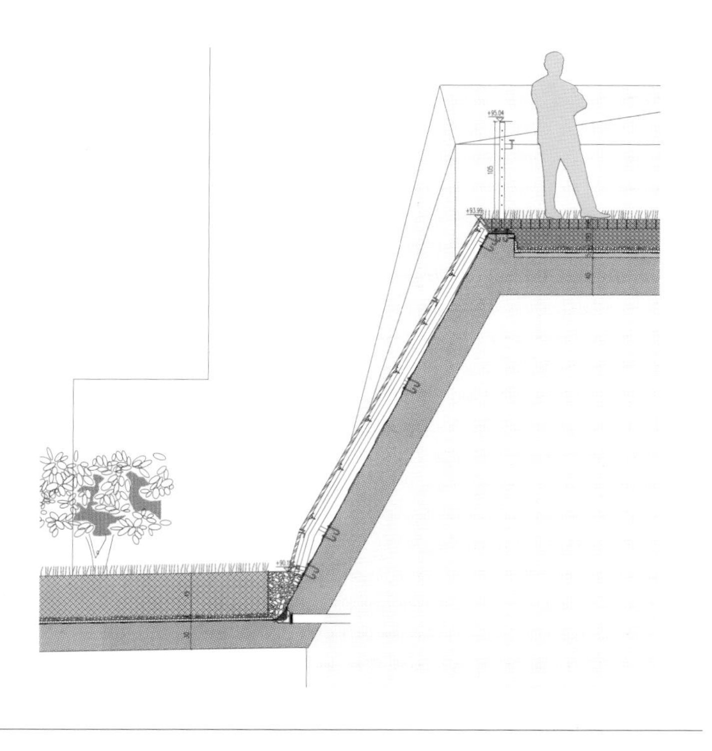

NEO 河岸公寓

NEO Bankside

设计公司： Gillespies

地点： UK（英国）

面积： 7 700 m^2

摄影： Gillespies/Jason Gairn

Developed as an integral part of the residential scheme, the landscapes at NEO Bankside designed by Gillespies provide richly-detailed green areas that balance beautifully with the contemporary apartment pavilions designed by architects RSHP. Unusually in the heart of a city, the new outdoor spaces offer NEO Bankside's residents opportunities to engage with nature.

The landscape designs take cues from natural processes found within woodlands, and transpose them to the city. A beehive is installed, and an orchard of fruiting trees and a herb garden give residents access to produce, and add colour and fragrance to the garden areas.

Introducing a high level of biodiversity, Gillespies worked with planting consultant Growth Industry, and incorporated large tracts of native plants set within groves of trees to provide a bank of flowers, seeds and nesting material to encourage wildlife to the space.

NEO Bankside's green spaces offer both residents and members of the public passing through a mix of the landscape typologies we find in nature. This approach creates a rich microcosm of landscapes within a constrained footprint.

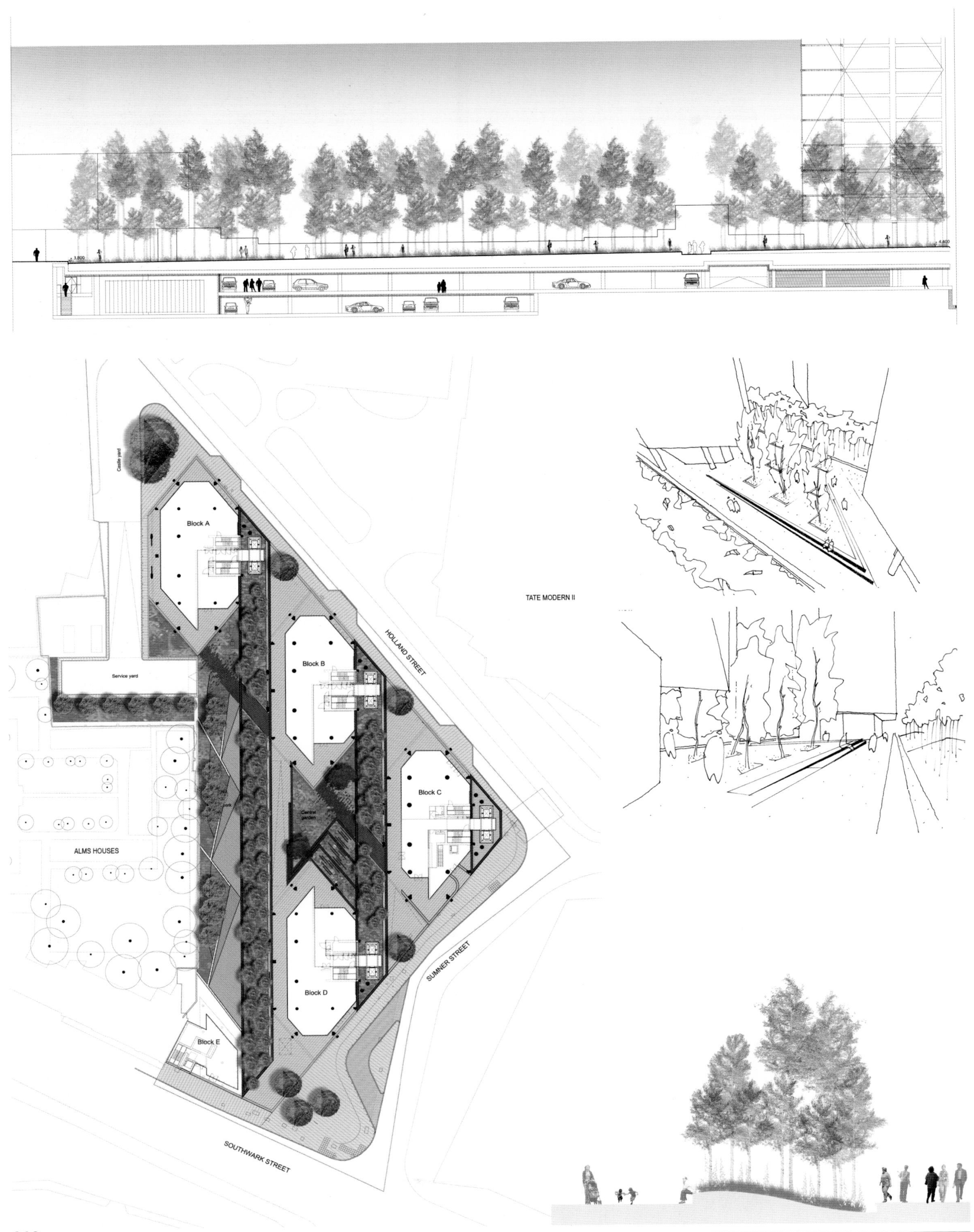
3.800
4.800
Castle yard
Block A
Block B
Block C
Block D
Block E
Service yard
Central garden
ALMS HOUSES
TATE MODERN II
HOLLAND STREET
SUMNER STREET
SOUTHWARK STREET

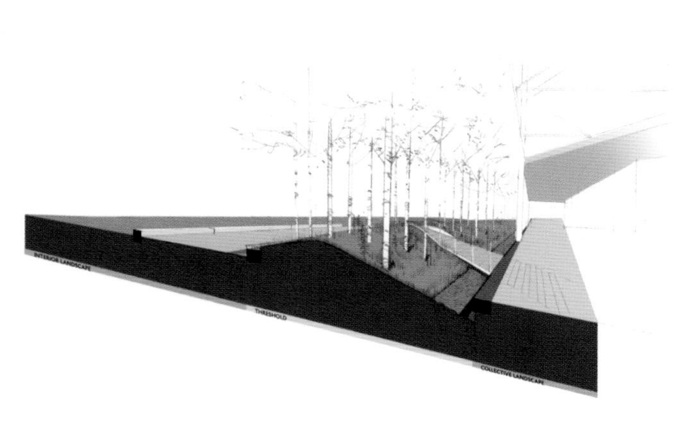
INTERIOR LANDSCAPE
THRESHOLD
COLLECTIVE LANDSCAPE

作为住宅计划的一个组成部分，NEO 河岸公寓拥有 Gillespies 景观事务所设计的细节丰富的绿地，完美协调了由 RSHP 建筑事务所设计的当代公寓。在城市的中心地带，全新的户外空间为住户们提供了难能可贵的接触自然的机会。

景观设计将林地自然枯荣的过程移植到城市中。园林中安置了蜂箱，果园和药草园带给住客一个多彩、芬芳的园区。

Gillespies 景观事务所与种植顾问合作，引入的物种保持了高水平的生物多样性，在大片区域内种植不同植被，以便形成开花、结果、筑巢等动植物的原生生态。

NEO 河畔公寓的绿色空间，向住客和途经的公众展示了能在自然界中找到的各种景观类型。这种丰富的微缩景观也有效避免了人们对自然的践踏。

太平洋坎纳公寓

Pacific Cannery Lofts

设计公司： Miller Company Landscape Architects

地点： USA（美国）

摄影： Dennis Letbetter

Pacific Cannery Lofts is constructed on a 1.09hm^2 site in Oakland's westernmost neighborhood. The project features the adaptive reuse of a historic cannery building from 1919 in a new residential community. The site design includes three internal garden courtyards: the Dining Room Court, the Living Room Court, and the Lew Hing Garden (named for the founder of the Pacific Canning Company. Hing was one of the nation's first Chinese-American industrialists.) Portions of the roof of the original cannery were removed to bring light to the new courtyards.

A rainwater harvesting system in the Dining Room Court channels water from the roof into raised concrete aqueducts with built-in bench seats. The rainwater system provides irrigation for courtyard plantings. A large communal dining table in the courtyard allows residents to come together for shared meals.

The Living Room Court includes centrally-located concrete banquettes and low tables that encourage gathering and conversation. Dogwood, brugmansia, and rhododendron create lush shade, highlighted by golden Hakone grass.

The Lew Hing Garden, the smallest of the courtyards, is an intimate space with a redwood boardwalk surrounded by Japanese maples, camellias, ferns, and coral bells.

The project design also incorporates linkages to surrounding sites, including two public pocket parks designed by Miller Company.

太平洋坎纳公寓位于，美国奥克兰西部的居民区，占地 1.09hm^2。该项目的特点是将 1919 年建设的一栋老建筑翻新改建成一片新的居住区。项目的设计包括 3 个内部花园庭院：餐厅式庭院、客厅式庭院和 Lew Hing 花园。原罐头厂的部分屋顶被拆除，使阳光能够进入新庭院。

一个雨水收集系统被安装在餐厅式庭院中，用来将屋顶上的雨水收集到混凝土水槽中。利用这个系统收集的水可以用来灌溉院子中的植物。院子里的大餐桌供人们在此处聚会、聊天或分享食物。

客厅式庭院中有一个位于中央位置的混凝土制成的座椅和一个矮桌子，是人们聚会聊天的好去处，这里种有山茱萸、曼陀罗、杜鹃花和箱根草。

Lew Hing 花园是最小的一座园子，这里有红衫木板路，还种植着日本枫树、茶花、蕨类植物和珊瑚钟。

考虑到了项目与周边环境衔接的问题，设计公司在项目周边配套建设了两个口袋公园。

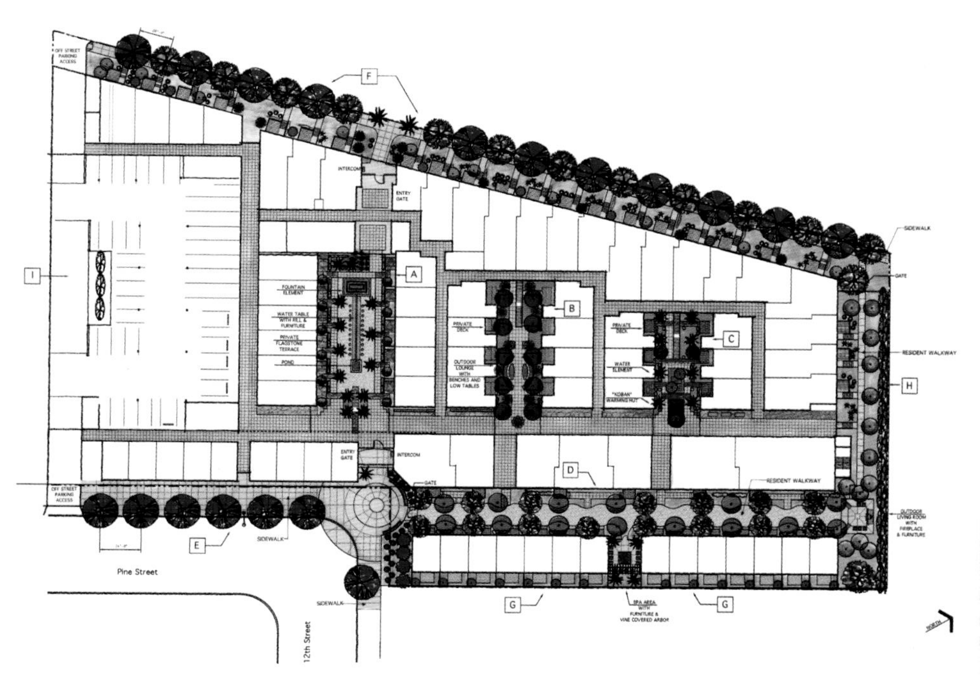
F
A
B
C
D
E
G
G
H
I
SIDEWALK
GATE
RESIDENT WALKWAY
INTERCOM
Pine Street
12th Street

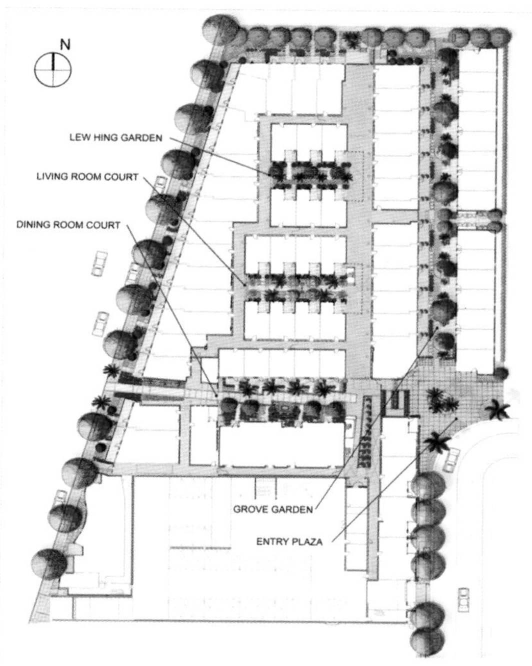
N
LEW HING GARDEN
LIVING ROOM COURT
DINING ROOM COURT
GROVE GARDEN
ENTRY PLAZA

Avalon 海洋大道

Avalon Ocean Avenue

设计公司： Miller Company Landscape Architects

地点：USA（美国）

摄影：Miller Company Landscape Architects

Avalon Ocean Avenue is a mixed-use residential and commercial development in the Balboa Park neighborhood of San Francisco.

This transit-oriented development is served by improved light rail and bus service along Ocean Avenue, as well as by the nearby Balboa Park BART station. The Ocean campus of the City College of San Francisco and the Ingleside branch of the San Francisco public library are immediately adjacent. Street-level retail includes a Whole Foods Market.

A pedestrian-oriented streetscape has been created along Ocean Avenue. Brighton Street extends into the development as an active "Woonerf" curbless street, with a raised terrace area that includes concrete seat walls and other pedestrian amenities. The street provides vehicles access to underground parking structures.

Residents can enter the building from the parking structure or through the street entry, which is flanked with palms in raised planters. Podium courtyards within the buildings create communal and private outdoor space for residents, with raised concrete planter boxes and a variety of fixed and movable seating.

The west podium features a wavy concrete bench and an outdoor cooking area with barbeques, a sink, and counters. Residents can enjoy a sunny day in the communal spaces, or sit in private patios on the ground level.

The east podium also allows for outdoor cooking and dining. It includes a dedicated space for gatherings with tables and chairs, as well as two

EXIT

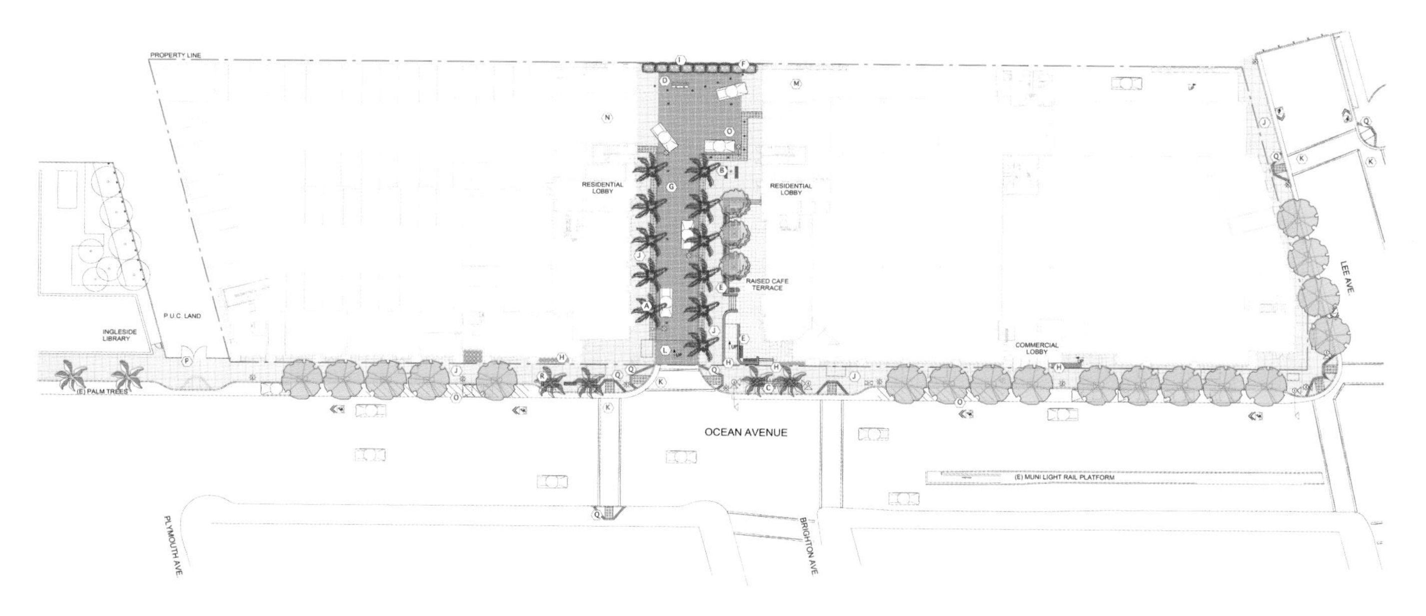
PROPERTY LINE
RESIDENTIAL LOBBY
RESIDENTIAL LOBBY
RAISED CAFE TERRACE
P.U.C. LAND
INGLESIDE LIBRARY
COMMERCIAL LOBBY
LEE AVE.
(E) PALM TREES
OCEAN AVENUE
(E) MUNI LIGHT RAIL PLATFORM
PLYMOUTH AVE.
BRIGHTON AVE.

EXIT

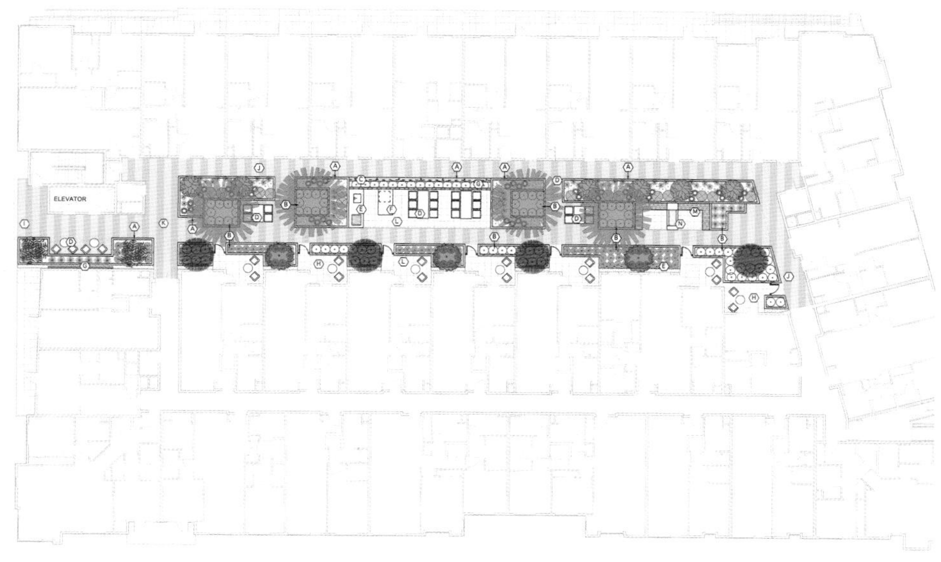

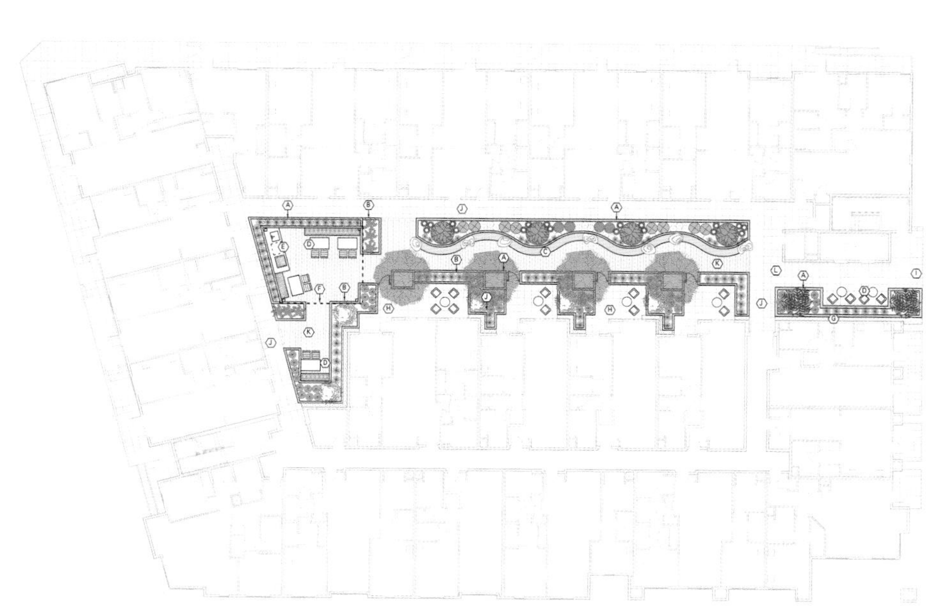

comfortable seating areas around a fireplace. Raised planters provide some privacy and enliven the central space.

Avalon 海洋大道为商住两用项目，位于旧金山巴尔博亚公园附近。

基于海洋大道沿途的轻轨、巴士及湾区交通服务，该项目以便利的公共交通为主要目标。旧金山城市学院海洋校区和旧金山公共图书馆英格尔赛德分馆与此相邻。商业街上有美国有机商品超市。

海洋大道的步行街与布莱顿大街相接，形成一个开放的住宅区，凸起的露台上有固定座椅等行人设施，地下建有停车场。

居民可以从停车场直接进入，亦可从种有棕榈树的街道走入。公共和私人的户外空间都位于群楼中的庭院里，那里有悬空种植的绿植及各种固定或移动的座位。

西裙楼设有波浪形的混凝土长椅和户外烹饪区。居民可在公共区域享受阳光，或在私人庭院闲坐小憩。东裙楼也可以在户外烹饪和用餐。室外不但提供了用于聚会的桌椅，围绕壁炉还有两个舒适的休息区。悬空绿植则把空间点缀得更加私密而灵动。

第三街区 5800 号

5800 Third Street

设计公司： Miller Company Landscape Architects

地点： USA（美国）

面积： 1.74hm^2

摄影： Miller Company Landscape Architects

5800 Third Street is a mixed-use development in the Bayview neighborhood in southeastern San Francisco. The project includes three phases of development on a 1.74 hm^2 site. The first phase depicted here was competed in 2010. The development includes 239 new below market rate residential units in two buildings along with 0.93hm^2 of commercial space on the ground floor facing Third Street.

The landscape design creates pedestrian relationships with the surrounding neighborhood and a variety of internal outdoor spaces for residents. Lush plantings of trees, shrubs, and grasses bring texture and color to the outdoor areas. Varieties of maple, dogwood, ornamental fruit trees, and palms provide shade and vertical interest. Flowering shrubs and perennials such as lavender, hydrangea and abutilon attract birds and butterflies and create lively natural habitats. Sedges, New Zealand flax, and grasses offer textural contrast.

Miller Company's design for the project faces the Carroll Avenue light rail stop of MUNI's F Line. The streetscape and plaza provides seating and greenery with a series of raised curvilinear planters and integrated seat walls. The planters and seat walls are constructed of concrete tinted a deep terra cotta red that contrasts with the concrete paving in broad strips of charcoal and gray. The seat walls are detailed with insets of glazed black tile. Palms and grasses in the planter beds create soft textures against the busy urban street.

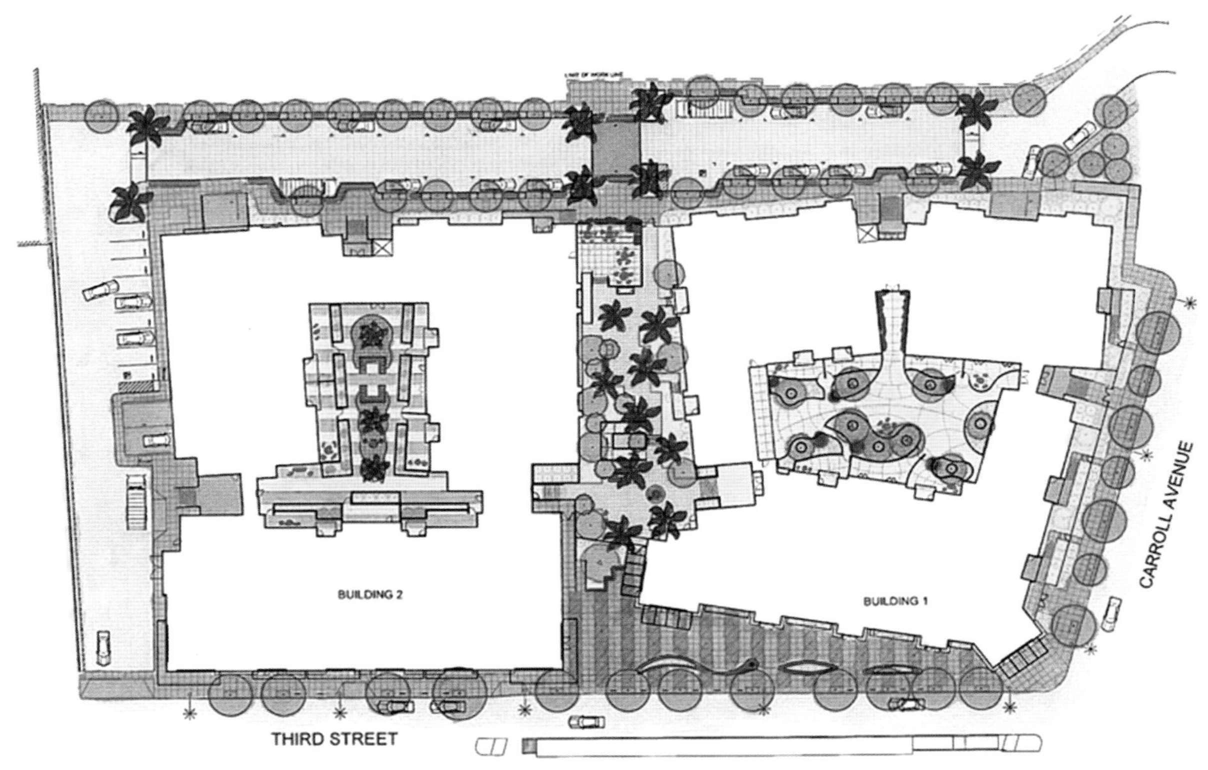
BUILDING 2
BUILDING 1
THIRD STREET
CARROLL AVENUE

Both buildings feature open-air central landscaped courtyards built on podium offering communal and private areas. Each courtyard features its own unique variety of geometries, materials, plantings, and seating opportunities. The north courtyard (Building 1) includes large curvilinear raised planter beds with concrete walls, accented by privacy fencing made of woven strips of galvanized steel. The south courtyard (Building 2) features central raised planting areas with serpentine borders, edged in galvanized steel. Perimeter privacy fencing is made with tinted translucent polycarbonate panels, accented by redwood and galvanized steel.

The mid-block pedestrian greenway links the two buildings to surrounding streets and provides public access through the development. The area is furnished with custom made redwood log benches and features a dog wash station for the convenience of the many resident dog owners.

第三街区 5800 号是位于旧金山东南湾景的一个综合项目，占地 1.74hm^2，分三期开发。两座建筑里共 239 个住宅单位，配有 0.93hm^2 的地面商铺。景观设计包含大量户外空间。郁郁葱葱的树木、灌木及草坪为室外区域平添了质感与色彩。种类繁多的枫树、山茱萸、观赏类果树及棕榈树遮蔽了烈日。开花灌木和其他多年生植物创造出自然栖息地，薰衣草、绣球花和茴麻等吸引着鸟类和蝴蝶。莎草、新西兰亚麻等草类各具特色。

Miller 公司负责设计的城市广场，正对着卡罗尔大道轻轨站。街道和广场上有座椅，绿化带蜿蜒其间。绿化与休息区被漆上了深砖红色，与灰色的混凝土路形成鲜明对比。座椅墙上的黑色琉璃瓦装饰匠心独具。棕榈树和青草的种植床给繁忙的城市街道带来柔和的肌理。

两座建筑都在裙楼顶部建有开放式中央景观庭院。每个庭院有自己独特的造型、材料、绿化和休息区。北部庭院（1 号楼）有混凝土砌筑的大型曲线形高架种植床，以不锈钢编织网隔离。南部庭院（2 号楼）中央种植区为蛇纹石边框、不锈钢镶边。周边隐私围栏采用有色半透明聚碳酸酯板，用红杉和不锈钢装饰。

中央街区的行人绿道连接着两座建筑物和周围街道。该区域配有定制的红木长椅，还贴心地为养狗的住户设置了洗狗站。

240 南方中央公园

240 Central Park South

设计公司： Balmori Associates, Inc.

地点：USA（美国）

面积：1 200 m^2

摄影：Mark J. Dye

The green roofs and entry courtyard of 240 Central Park South pull the character of Central Park through the building and up to the roof. Contoured ribbons of shrubs and sedums are interwoven with lines of slate, mimicking the rock outcroppings in the park.

This landscape is designed to be experienced from multiple viewpoints. Visitors walking by the building catch glimpses of the cherry trees peaking over the parapet wall, while tenets inside the building are surrounded by the rolling ribbons of plants. From the neighboring buildings and apartments above, the multiple levels of rooftops appear to join together into one unified landscape.

240 南方中央公园入口处的庭院和绿色植物装饰的屋顶体现了项目的特色。设计独特的带状灌木丛和景天植物丛与石板路交织在一起，形成了独特的景观。

这一景观的设计为游客提供了多个观赏角度。游客可以看见越过挡土墙的樱花树，在内部可以观赏到建筑楼宇间呈带状分布的植物。从附近的建筑和公寓，人们都可以欣赏到不同层次的景观。

巴约纳的阿尔塔夫拉别墅

La Torreta de Bayona

设计师：David Jimenez, Jaime Pastor

地点：Spain（西班牙）

面积：8 500 m^2

摄影：Patricia Reija, David Jiménez

On the north of Alicante, we find this villa built on the sixteenth century, The landscape is arid and rugged, typical of southern Spain where the land is white and dry.

In many parts of the Spanish Southeastern, irrigation techniques still remain from the Arab period, such as the use of irrigation channels. This technique is still used on the fields of La Torreta de Bayona to build a new irrigation pond was imperative for the program's project. This need, along with the fact of giving meaning to the garden, result in a landscape project where the oasis is a source of life. The rugged terrain is transformed into a green landscape with reflections and shades. An area full of dreamscape places which protect us from dry terrain.

The project is based therefore on the construction of a lake which also used as irrigation store. The lake is the heart of this large Mediterranean garden with oasis soul. It has over 2,100 cubic meters and an average depth of 120cm. Its boundaries are blurred by the use of grasses as Pennisetum rubrum, Stipa tenuissima. Besides, the Juniperus pfitzeriana contrasts with grass.

The lake is conceived as a space to be seen and discovered, especially the carob tree which, like a living sculpture, presides over one of the two islands. A path, sometimes width and natural and other tight and hard, runs the garden. The tree together with stones, make difficult the vision of the lake. A vision that changes depending on the season, being this is a highly seasonal garden.

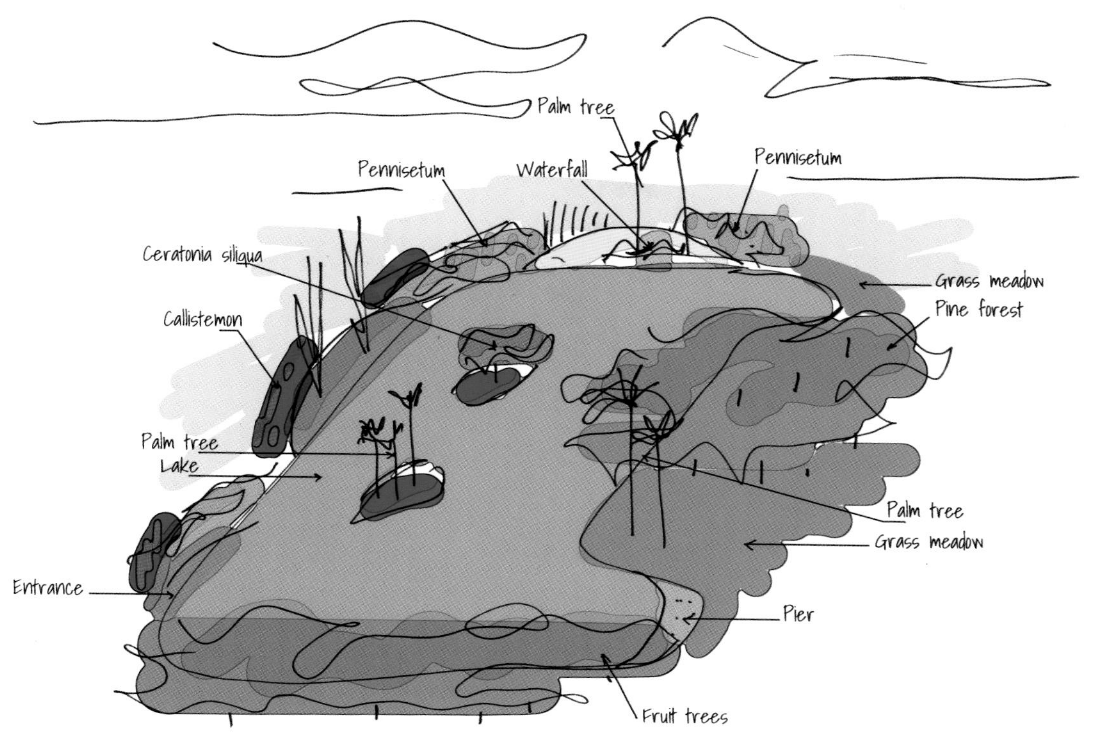

Palm tree
Pennisetum
Waterfall
Pennisetum
Ceratonia siliqua
Grass meadow
Pine forest
Callistemon
Palm tree
Lake
Palm tree
Grass meadow
Entrance
Pier
Fruit trees

The garden, although implanted in a harsh environment, uses a typical vegetation of the Mediterranean climate and therefore will evolve diluting itself with the landscape, making this landscaping project, part of the territory. The pine forest instead, filters the light creating an intimate atmosphere. It works as a hinge between the wild and natural lake area and celebrations meadow. The pines, like small columns support a vegetable roof; resulting in a landscape to explore and observe.

这座别墅建于16世纪，位于阿利坎特的北部。此处土地干旱，道路崎岖，土质是典型的西班牙南部的白干土。

在西班牙东南部的大片地区，仍广泛使用灌溉渠技术。如今，巴约纳的阿尔塔夫拉别墅花园也使用这种老式的灌溉技术，项目的当务之急是采用新的灌溉技术。改造后的花园是一个能够展示出生命力的沙漠绿洲。

该项目中有一个人工湖。这个湖是绿洲的灵魂，是地中海花园的心脏。它的蓄水量超过2 100m^3，平均深度120cm，周围种植了狼尾草、细茎针茅。此外，刺柏与草形成了强烈的对比。

湖的位置十分显眼，大乔木像活的雕塑般遍布整个花园。一条时宽时窄的小路直通花园。植物和石头的精心布置，创造出人工湖美丽的景致。这是一个季节性很强的花园。

虽然花园地处一个较恶劣的环境中，但地中海植被的大量使用使它成功地融入整个大环境中。松树林遮挡了多余的阳光，营造出温馨的氛围。它在湖泊和草甸之间自由地生长。松树林像是一个个支撑屋顶的柱子，激发了人们探索其中空间的欲望。

北贝尔瓦尔——居住景观设计

Belval North, 'Living (in) the Landscape'

设计公司： ELYPS Landscape + Urban Design

地点： Luxemburg（卢森堡）

面积： 1.9 hm^2

摄影： Hanns Joosten

A smoothly turning road is the backbone of the residential area, showing various perspectives on buildings and landscape.

The architecture and internal design from the mixed apartment blocks and row of houses strive to a maximum contact with the landscape. The courts are open towards as well road as the landscape. Balconies and loggia's offer a splendid view.

The gardens are limited in depth (5-7m) so the contact with the landscape is not lost, and have 1.2m high hedges. This guarantees the inhabitants privacy as well as a view on the landscape. All gardens have access to the green wedges in between the buildings, thus preventing it to be anonymous residual spaces.

The spaces between the buildings and the street are just modestly arranged, so the boundary between public and private is not to sharp and so strengthens the concept of buildings on a 'green carpet'.

The green carpet carries the design of the whole Belval North project. The number of ingredients used in the landscape design is modest, so the design does not compete with the architecture. Hedges, height differences, ditches, mowed grass strips, islands of wild flowers and trees scattered are sufficient in creating a clear and recognizable basis for variations in architecture of the building blocks.

The design and abstract refinement of the landscape is functional

KURT
SOLUDEC
KURT
TECHNOCONSULT
TRACOL
ASARS
C BELVAL
Esch-sur-

and straightforward. Spaces for the natural development of domestic vegetation alternate with tightly mowed grass strips one can walk or let the children play on.

有一条岔路可以直接到达这片居住区，路边的房屋和景色优美怡人。
建筑的内部设计追求最大化地与景观构成联系。院子朝向道路开放，在阳台上可以欣赏到广阔的景色。
花园纵深 5~7m，挡土墙高 1.2m，保证了居住者的个人隐私和观赏高度。
建筑之间是绿色的种植，为单一的居住环境增添了色彩。
人居建筑和公共建筑之间的界限并不明显，加强了“绿色地毯”的建筑概念。
北贝尔瓦尔项目整体在绿色的“地毯”上建成。设计师用一些元素——树篱、沟渠、修剪草带、野花、绿树恰到好处地创造了一个清晰美观、辨识度高的建筑空间。
景观设计讲究实用性和直观性。孩子们可以在自然植被、草甸上尽情玩耍。

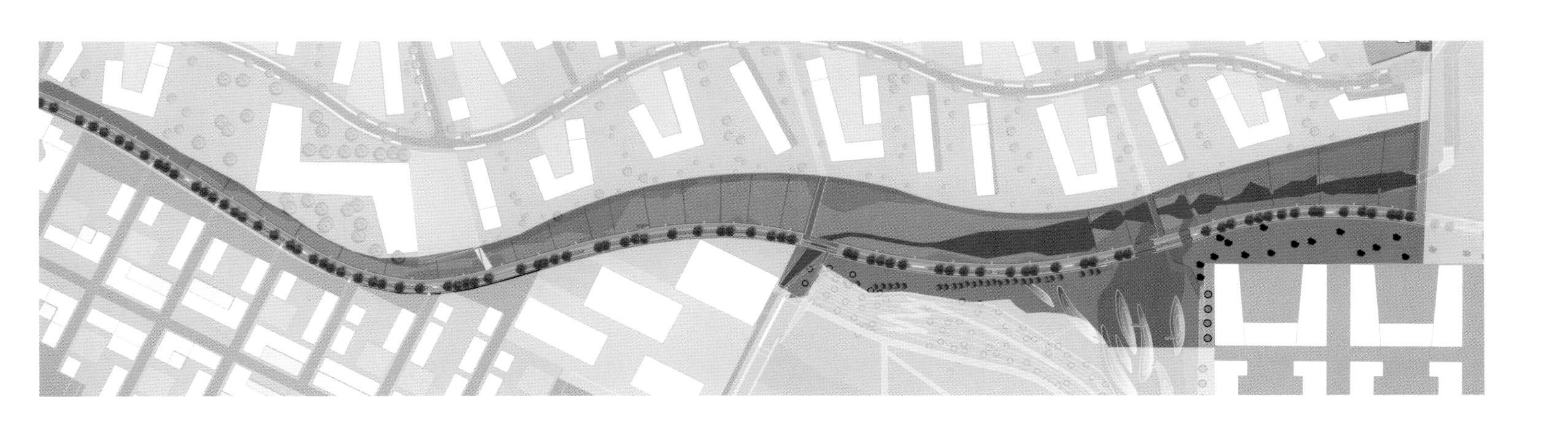

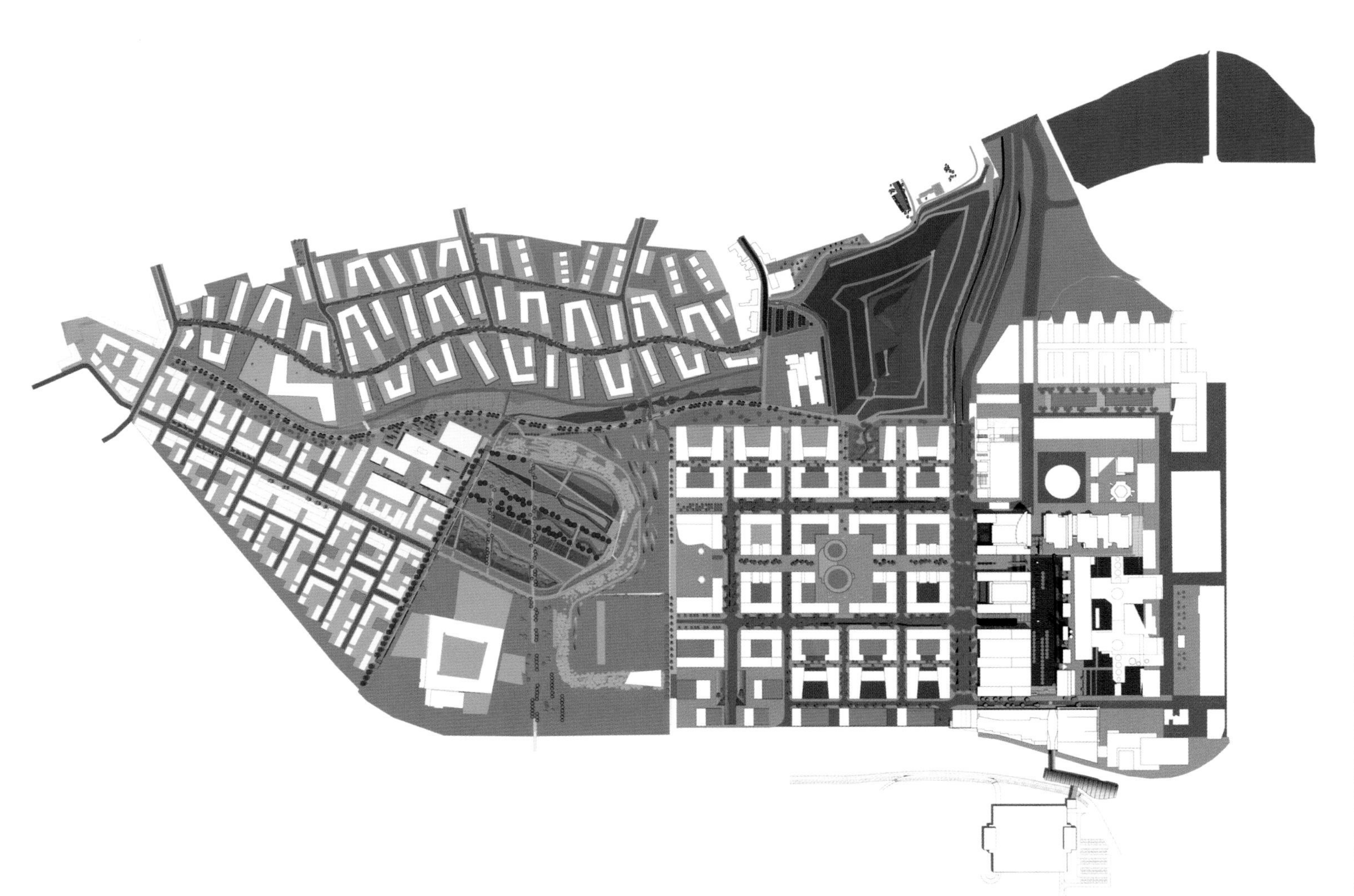

WT12

[13]

[13]

格罗夫纳花园

Grosvenor Park

设计公司：Landscape Architecture Bureau

地点：USA（美国）

面积：1011 m^2

摄影：Prakash Patel

The Grosvenor Park Condominium commissioned Landscape Architecture Bureau to redesign two roof gardens, each approximately 1,000 square meters, one on each side of the condominium's lobby. The need for the project came about as a result of the aging of the building, constructed in 1963. The waterproof membrane between the roof garden and the parking garage below had failed. Leaks had damaged cars and caused corrosion to the structural steel of the concrete roof slab. In order to repair the slab and replace the waterproof membrane, all the existing soil, plants and paving had to be removed. LAB was charged with designing the new "replacement" gardens.

The new gardens are experienced both at ground level and by views from above. A series of parterres of low shrubs and perennials (Stachys Byzantium, Juniperus horizontalis, Hypericum calycinum, Helleborus niger, Hemerocalis 'Stella Doro', Nepeta x faassenii) provide visual interest. In each parterre is a single, specimen Honeylocust (Gleditsia "Shademaster") is planted, forming a line leading to a shaded seating area. In addition, an existing terrace on each side, adjacent to the lobby, was made smaller and repaved to reduce an overlarge hardscape and to make these areas more inviting. A critical issue to the residents was the need for protection from prying eyes for the ground floor units. Extensive screen planting was installed to maintain a sense of security and privacy for the

ground floor residents and at the same time to invite residents to inhabit the new, communal outdoor spaces.

格罗夫纳花园委托景观设计师对花园进行重新设计。花园占地大约 1 000m^2，紧邻公寓入口处。建筑始建于 1963 年，很多设施已经老化，地面和地下停车场之间的防水层已经破损。积水损坏了车辆，腐蚀了停车场混凝土屋顶的钢结构。如果要修复这些设施，必须先移开现有的植被和石板路。因此，LAB 设计事务所可以重新设计花园了。

新的花园可以从地面或高处观赏。一系列低矮的灌木和多年生植物（拜占庭水苏、杜松、大萼金丝桃、黑嚏根草、斯特拉多罗萱草、荆芥）为花园提供了美丽的视觉景观。设计师采用了规则式的形式种植了乔木和地被，形成了线形景观。另外，入口两侧的露台面积被缩小，并且重新进行铺装设计，从而弱化了硬质景观。大片植被代替了硬质景观，有效地保证了一层居民的隐私，同时有利于住户们共同维护这处花园。

阿姆斯特尔芬关怀中心

Amstelveen Zonnehuis Care Home

设计公司：HOSPER

设计团队：Ronald Bron, Frits van Loon, Elizabeth Keller, Petrouschka Thumann, Marike Oudijk

地点：the Netherlands（荷兰）

面积：3 hm^2

摄影：Pieter Kers, HOSPER

Town dwellers need high-quality green space to enhance their living experience. HOSPER has produced a proposal for development of the grounds of the Zonnehuis Care Home in Amstelveen into a lively green pedestrianized area. The various buildings on the site are linked by a square paved in a distinctive design housing a number of richly planted green oases and play and other facilities. Four residential units on a multifunctional plinth, designed by Rijnboutt, will be erected on the west side. A water-rich patio garden has been designed as part of this complex.

The new Zonnehuis care complex, designed by Architectuurcentrale Thijs Asselbergs, has already been completed on the east side. The outside space surrounding the care complex consists of gardens with perennial plants and a greenhouse where residents can garden. The square includes a terrace and a play area, a garden area with perennial plants and a sloping path and steps for walking exercises. Two gardens have been specially designed to be dementia-friendly, with the aid of advice from Anke Wijnja of Bureau Fonkel and Annie Pollock of the Dementia Services Development Centre at the University of Stirling in Scotland.

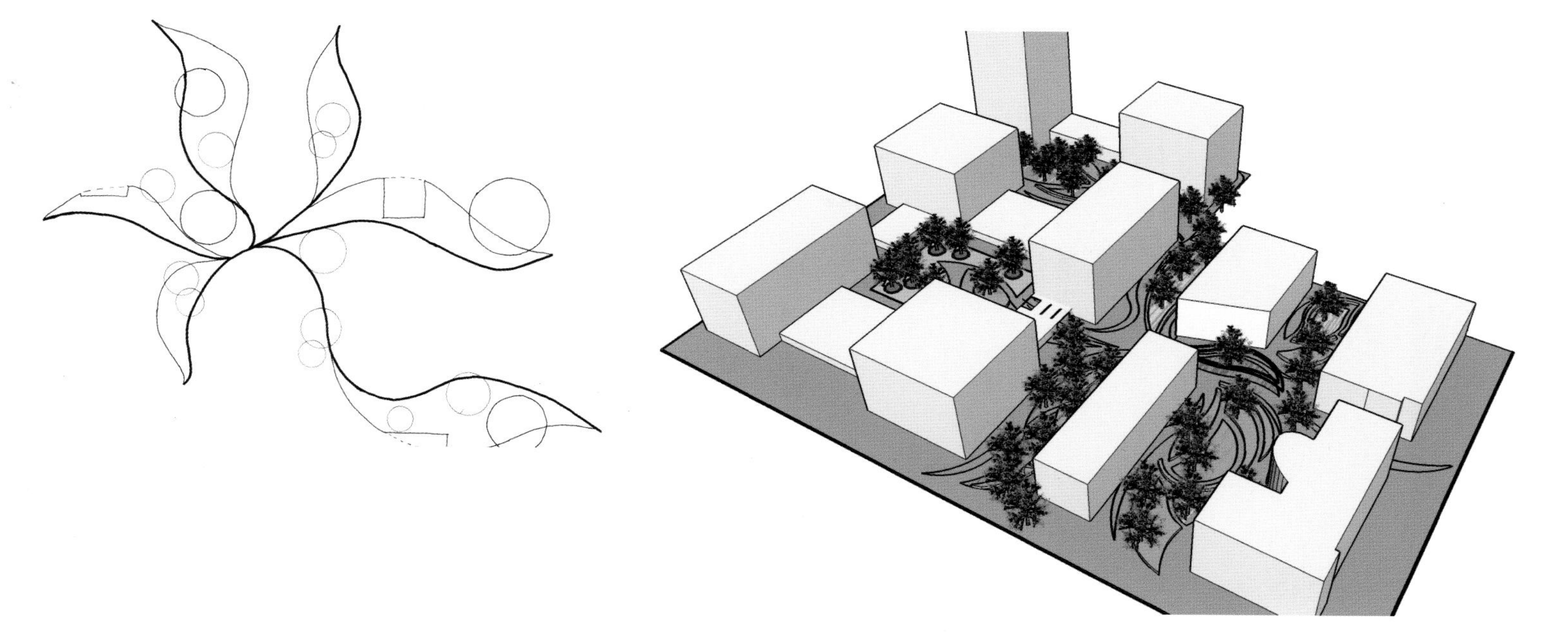

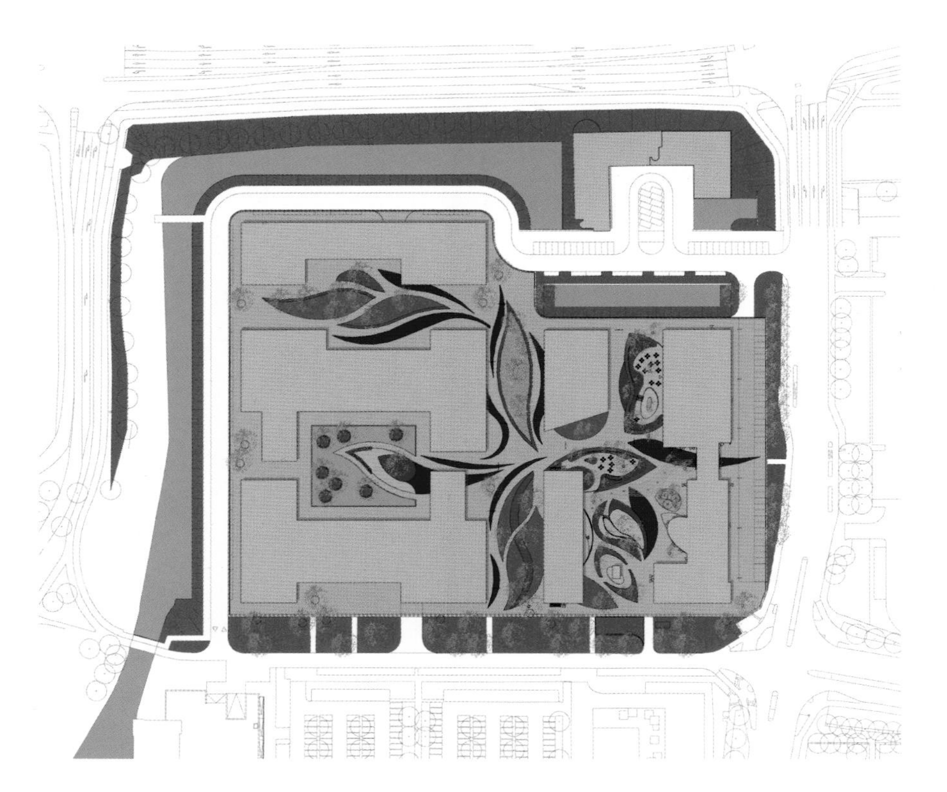

荷兰阿姆斯特尔芬的小镇居民需要高质量的绿色空间，以优化他们的居住环境。设计师提出了一种设计方案，即在绿色步行区建造阿姆斯特尔芬关怀中心。项目的多个建筑之间由一个广场连接。项目的设计十分独特，周边种植了丰富的植物，并设置了娱乐设施和多种其他设施。水源充沛的水景花园也是建筑项目的一部分。

关怀中心项目中建筑的东面已经建设完成。项目的外围包括花园、多年生植物、居民自己的温室。广场则由梯田、游乐场、多年生植物花园区、锻炼身体的斜坡小路组成。根据学者 Anke Wijnja 和斯特林大学老年痴呆服务发展中心工作人员的建议，设计师特别将其中两个花园的景观打造成适合患有老年痴呆症的人群休憩的地方。

100+
LANDSCAPE
DESIGN

教育
EDUCATION

新加坡科技与设计大学

Singapore University of Technology & Design

设计公司： UNStudio, DP Architects

地点： Singapore（新加坡）

面积： site 83 000 m^2

摄影： Hufton+Crow/Sergio Pirrone

Located close to both Changi airport - Singapore's principal airport - and the Changi Business Park, the SUTD is Singapore's fourth public university.

The design for the SUTD campus aims to achieve the highest Green Mark rating (platinum) available in Singapore. Preliminary considerations in the design include building orientation and depth in relation to sun and wind exposure, along with the incorporation of maximum natural ventilation and daylight to all buildings.

The block orientations are configured to minimise East/West solar exposure with considerations of inter-block shading, while a porous ground floor enhances updrafts. The well shaded and well lit courtyards are connected to campus wide circulation spaces through 'wind corridors' that direct Northeast and Southeast prevailing winds into the courtyards.

Air conditioned spaces are reduced by means of naturally ventilated perimeter corridors that create a shading overhang and minimise heat loads. Horizontal louvers are designed to reduce solar gain, reflect and diffuse daylight entering internal spaces and shield from tropical rains on exterior corridors. The coloured precast, aluminum and glass façade increases daylight where needed and allows for a flexible integration of M&E louvers.

The façade of the academic campus responds to its urban location, while

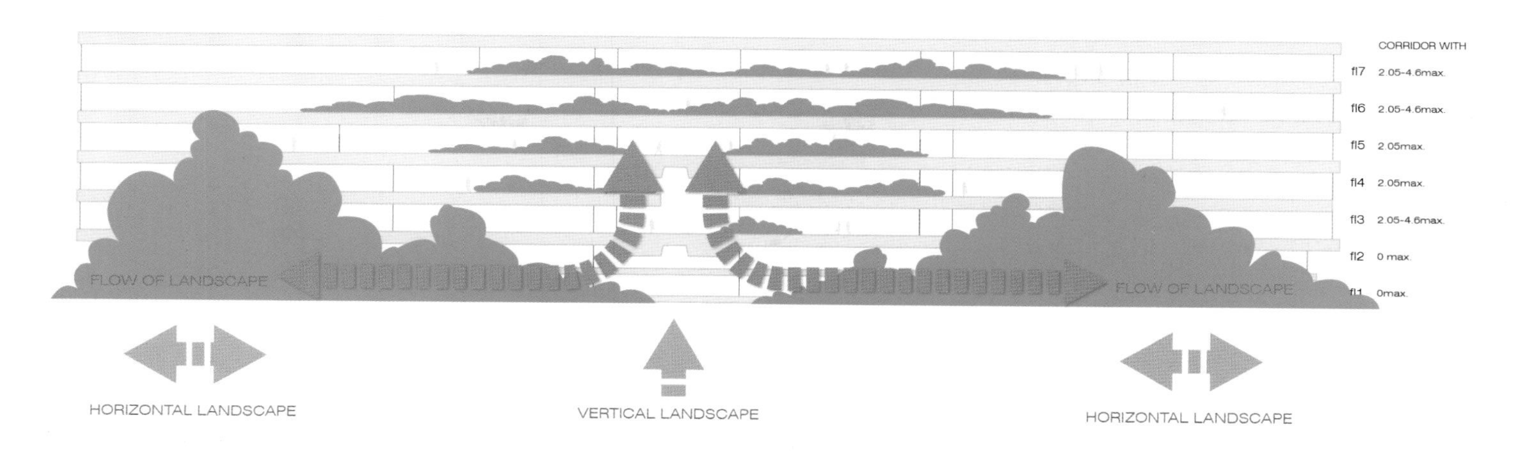
CORRIDOR WITH
fl7 2.05-4.6max.
fl6 2.05-4.6max.
fl5 2.05max.
fl4 2.05max.
fl3 2.05-4.6max.
fl2 0 max.
fl1 0max.
FLOW OF LANDSCAPE
FLOW OF LANDSCAPE
HORIZONTAL LANDSCAPE
VERTICAL LANDSCAPE
HORIZONTAL LANDSCAPE

tree-shaded walkways and basement planters generate cooled, outdoor modes of circulation that render the campus walkable and low-carbon.

新加坡科技与设计大学临近新加坡樟宜机场和樟宜商业园，该大学是新加坡第四所公立大学。

新加坡科技与设计大学校园设计获得新加坡建筑行业的环保最高奖等级——绿色标志评级（白金级）。设计的初步考虑是建筑的朝向和高度，这与采光和通风都有关系，设计目标是争取每栋建筑都能将采光和通风最大化。

设计充分考虑室内环境、建筑的朝向设计，地板的设计增强了建筑的通风性能。阳光与阴凉都具备的后院通过通风走廊与校园的圆形场地连通，通风走廊将南北向的风引入建筑。

自然的通风走廊能够遮挡阳光、降低热负荷，从而减少空调的使用。百叶窗的设计减少了多余阳光进入室内，同时遮挡户外的雨水。必要时，也可以调整彩色的预制铝玻璃外墙和百叶窗来增加日光照射。

学术校园的外墙设计十分时尚、现代，符合新加坡的城市形象，林荫路和苗圃种植园让人更亲近自然，户外的景观设计使校园环境更生态、舒适。

SUTD

AUDITORIUM
CAMPUS CENTRE
ONE STOP CENTRE
CAMPUS CENTRE
ONE STOP CENTRE
47
49

The Hive 公共图书馆

The Hive

设计公司：Grant Associates, Feilden Clegg Bradley Studios

地点：UK（英国）

面积：2 hm^2

摄影：Grant Associates

The Hive is Europe's first joint university and public library – a unique academic, educational and learning centre for the City of Worcester and its University. The 'BREEAM Outstanding' project was designed by architects Feilden Clegg Bradley Studios with a distinctive and sustainable landscape design by Grant Associates.

Grant Associates' landscape design brief was to create a high quality landscape environment that would become a distinctive and exciting visitor attraction - a place which would capture a sense of history and place whilst reflecting on the contemporary themes of sustainability and technological innovation.

The landscape is based on a strong narrative derived from the local landscape of the River Severn, Malvern Hills and the Elgar trail that inspired Land of Hope and Glory and key storytelling themes:

Nature uplifts the spirits The landscape spaces are arranged to 'enlighten and delight' inviting visitors to experience the therapeutic qualities of nature, an encounter with birdsong, scented plants, colourful wildflowers and dragonflies.

Healthy water for sustained life Demonstrates to visitors the importance of healthy water for life and the ability of natural systems, not man made chemicals, to take care of this.

Knowledge and Heritage Creates a special sense of place derived from the primary circulation route The Causeway.

BUS
LANE

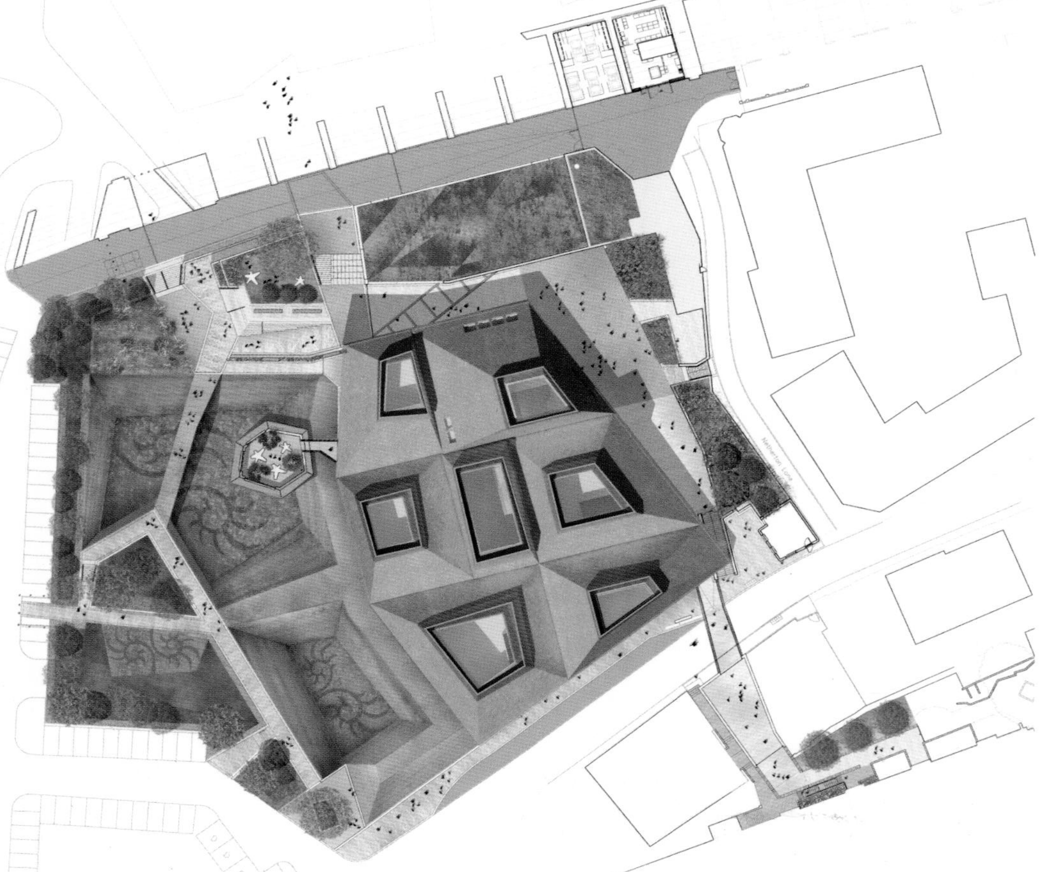

The two-hectare site comprises a series of islands and belvederes overlooking two landform basins containing rich local damp meadow and The Causeway, an extension of the city wall route, routes circling around and through the Centre. Highlights include.

The Hive 是欧洲第一联合大学公共图书馆，也是伍斯特大学城的学术、教育和学习中心。设计师设计的可持续景观风格鲜明，获得了“BREEAM 杰出项目”认可。

景观提案旨在建造一个引人入胜的高品质景观环境，在反映当地历史和本土性的同时，展现出可持续发展与技术创新的面貌。

景观叙事以当地的塞文河和莫尔山为出发点，围绕希望与荣耀展开，主题如下。

自然振奋精神 景观空间营造出鸟语花香的氛围，以“启迪和愉悦”引领游客体验大自然的疗愈之功。

滋养生命的健康之水 向游客展示健康水生活的重要性，尤其是通过自然系统，而不是人造化学物质来实现这一切。

知识和遗产 铺设长堤，形成一个特别的主循环路线，占地 8 094m^2，有一系列岛屿和观景平台，可俯视两块盆地，潮湿的草甸和长堤蜿蜒其间，道路四通八达。

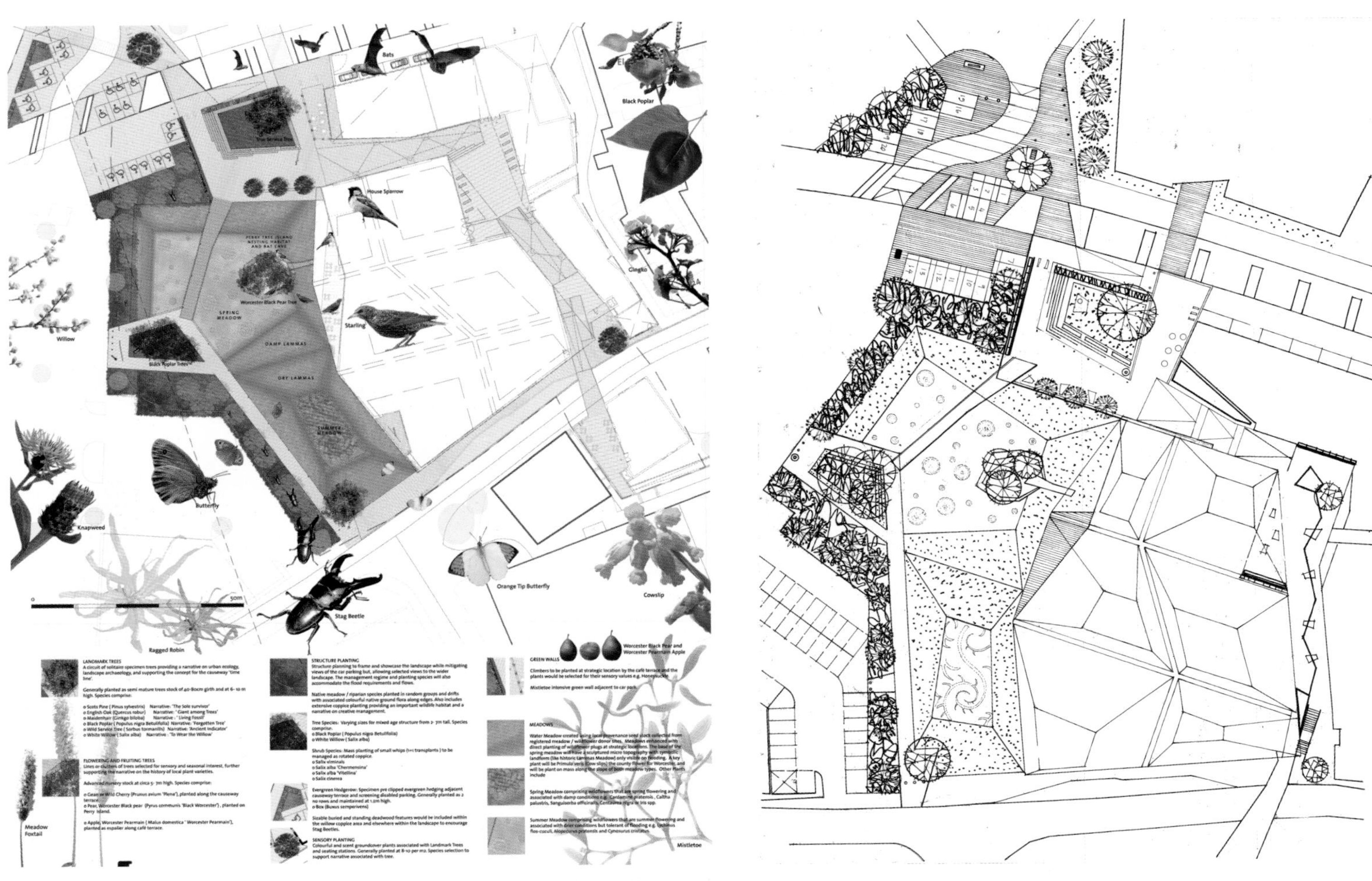
Bats
Black Poplar
House Sparrow
Ginkgo
Willow
Starling
Knapweed
Butterfly
Stag Beetle
Ragged Robin
Orange Tip Butterfly
Cowslip
Meadow Foxtail
Mistletoe
50m

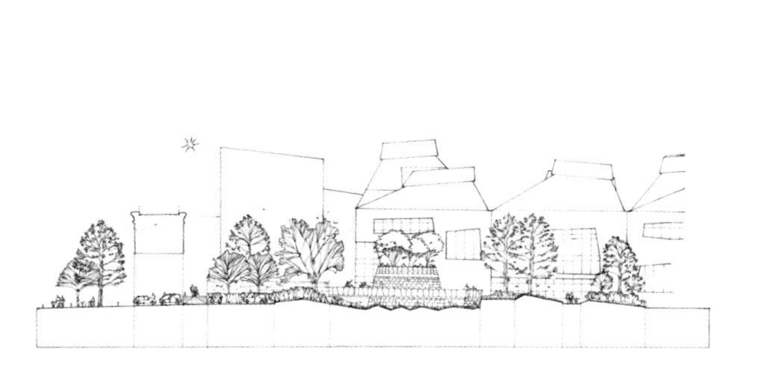

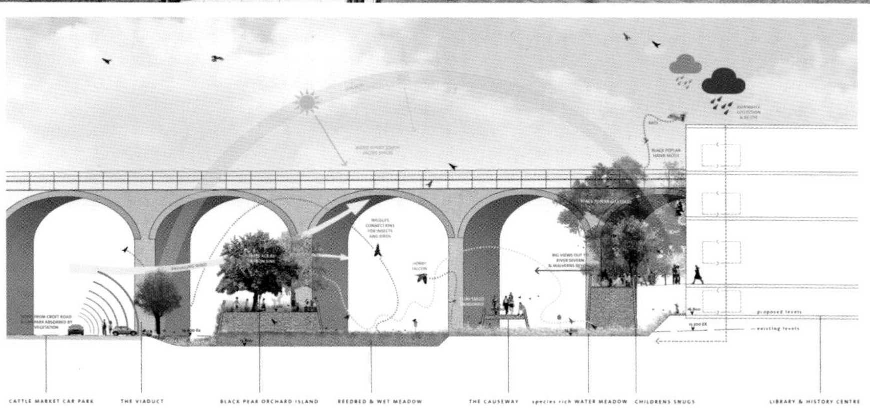
CATTLE MARKET CAR PARK
THE VIADUCT
BLACK PEAR ORCHARD ISLAND
REEDBED & WET MEADOW
THE CAUSEWAY
species rich WATER MEADOW
CHILDRENS SNUGS
LIBRARY & HISTORY CENTRE
proposed levels
existing levels

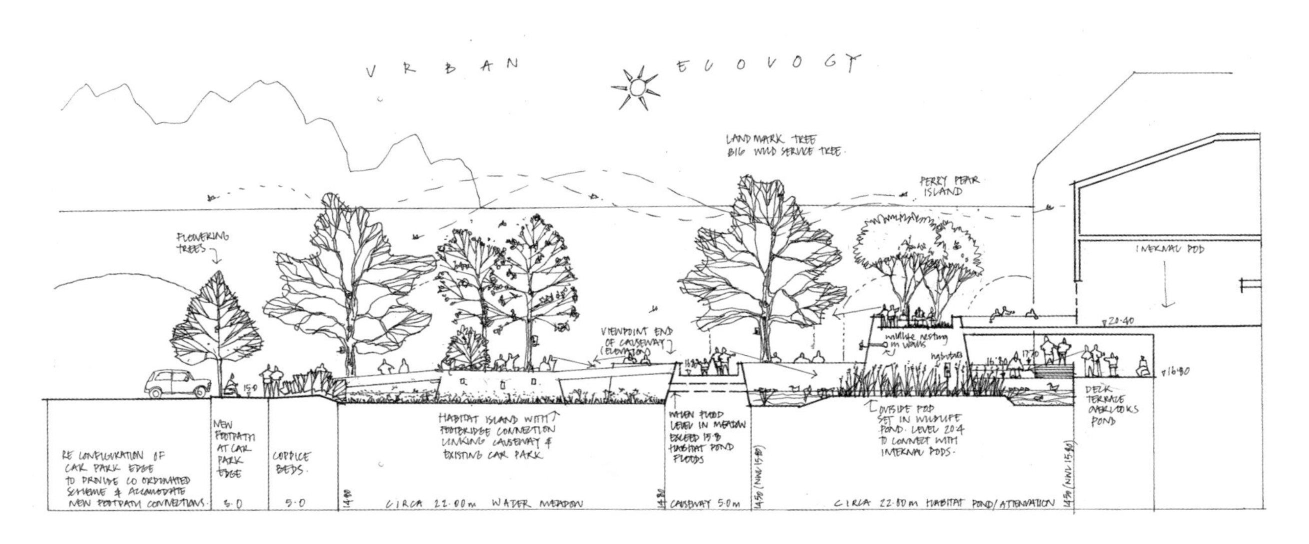
URBAN ECOLOGY
LANDMARK TREE
BIG WILD SERVICE TREE
PERRY PEAR ISLAND
FLOWERING TREES
INTERNAL POD
VIEWPOINT END OF CAUSEWAY (ELEVATED)
HABITAT ISLAND WITH FOOTBRIDGE CONNECTION LINKING CAUSEWAY & EXISTING CAR PARK
WHEN FLOOD LEVEL IN MEADOW EXCEED 15.5 HABITAT POND FLOODS
OUTSIDE POD SET IN WILDLIFE POND. LEVEL 20.4 TO CONNECT WITH INTERNAL PODS.
DECK TERRACE OVERLOOKS POND
RE CONFIGURATION OF CAR PARK EDGE TO PROVIDE CO ORDINATED SCHEME & ACCOMMODATE NEW FOOTPATH CONNECTIONS
NEW FOOTPATH AT CAR PARK EDGE
COPPICE BEDS.
CIRCA 22.00m WATER MEADOW
CAUSEWAY 5.0m
CIRCA 22.80m HABITAT POND/ATTENUATION

圣爱德华大学蒙迪图书馆

St. Edward's University Munday Library

设计公司：Sasaki

地点：USA（美国）

面积： 4359 m^2

摄影：Casey Dunn

Defying predictions of its impending obsolescence, the academic library is undergoing a dramatic rebirth and reinvention. The continuing evolution of technology, the reassessment of the basic tenets underlying teaching and learning, and the ever increasing awareness of fiscal limits have driven a dramatic rethinking of the library's role on campus. This context of change prompted Saint Edward's University to dramatically transform its library. The Munday Library—an addition to and renovation of the campus's existing library—is imagined as a single, central space that enhances and catalyzes interaction around technology and group learning. The new library also provides a signature academic space that embodies the values of the Congregation of the Holy Cross—intrinsic connection to the community and grounding in the natural world.

Sasaki's pragmatic approach appropriately focuses attention and resources where they will have the most impact. Within the library, student interactions, research, and inquiry all happen within sight of one another and are supported by a technology-rich environment in a variety of study spaces. All student services are organized in the main commons and a single reference desk offers students a clear source of help with research, digital media, and reserve materials. The central location of the commons makes it a catalyst for all the programs and initiatives in the building. The commons is flanked by two classrooms, which are linked via IT infrastructure to St. Edward's sister campuses in Angers, France and Vina del Mar, Chile. The general collections

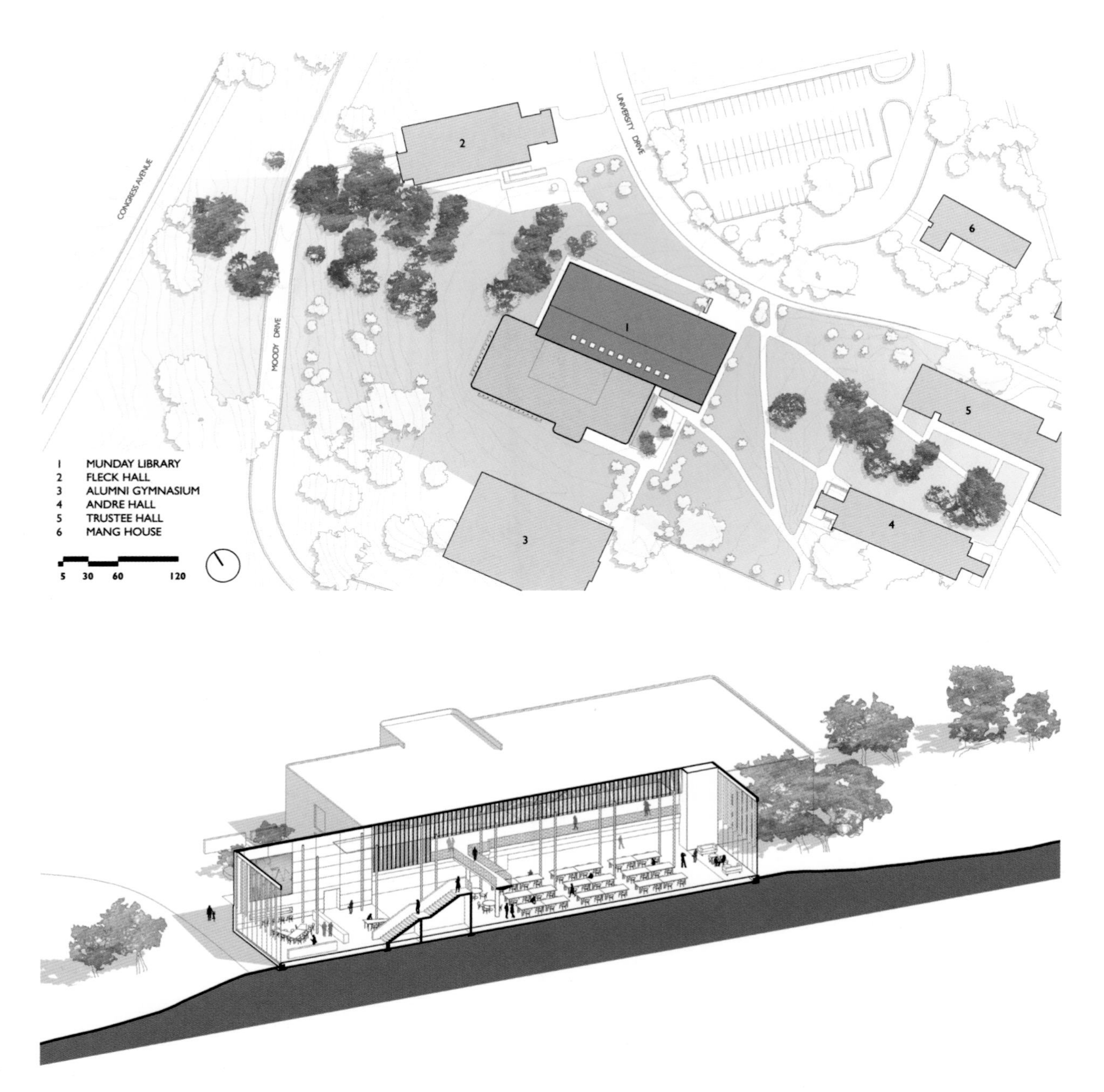

and the Writing and Media Center are on the second floor. The second floor also features a bridge that visually connects the two floors. An access flooring system delivers power anywhere it is needed now or in the future.

The Munday Library also reconnects a neglected grove of live oaks to the network of shady courtyards that dot the campus hilltop. Framed views at either end of the building emphasize these connections and establish the library as the academic heart of the university. Activity inside is highly visible upon arrival and late into the night when the library becomes a beacon glimpsed from paths in all directions. A courtyard situated at the building entrance provides shaded relief and acts as an extension of the commons with casual seating and outdoor study space.

为了逃避废弃的命运，学术图书馆必须经历彻底的重生和革新。在科技持续变革、教学基本原则重新评估，以及财力越来越局限的情况下，图书馆在校园中的角色引发了人们的思考。这些情形的变化催生了圣爱德华大学图书馆的巨变。蒙迪图书馆是对现有校园图书馆的新增和改造，它被定位为集中的核心空间，加强并促进技术方面的互动和小组学习。新的图书馆也提供了标志性的学术空间，它体现了圣十字教会的价值观——联系社区与扎根自然。

Sasaki 务实的设计将吸引点和资源策略性地安置在影响最大的地方。图书馆里，学生们在多个学习空间及各类高科技的支持下，进行互动、研究和答疑解惑。所有的学生服务都布置在主要的公共空间，唯一的咨询台向大家提供清晰的关于研究、数字媒体与备份材料的帮助。位于中央的公共空间成为建筑中所有功能和活动的催化剂。公共空间两侧各有一个教室，通过信息科技设施与圣爱德华在法国昂热和智利维纳德的兄弟校园相连。一般馆藏和写作与媒体中心在二层。二层以桥梁为特色，从视觉上与一层相连。地板供电系统将电力输送到馆内任何需要的地方。

蒙迪图书馆将一片被遗忘的橡树林重新纳入了校园山顶上的林荫内院。建筑周边的景致突出融合的特点，彰显了图书馆在大学学术核心的地位。人们一到校园就能看到建筑里面的情景，晚间的图书馆则成为从各个方向都令人瞩目的灯塔。建筑入口的内院有遮荫的区域，这也作为公共空间的延伸，提供休闲和户外学习空间。

布罗斯纺织学校时尚中心

Borås Textile Fashion Center

设计公司：Thorbjörn Andersson with Sweco architects

设计团队：PeGe Hillinge, Staffan Sundström, Ronny Brox, Per Johansson (lighting design)

地点：Sweden（瑞典）

In front of the entrance of the recently opened University of Textile & Fashion in Borås, Sweden, there is a carpet of stone, 110 meters long and 11 meters wide. The pattern of the carpet is inspired by a weaving method for complicated patterns; the so-called Jacquard technique. The carpet is laid out in three different types of granite and forms the hallmark for this new school and its campus. It is also the main public space for the school, openly programmed with the possibility of a lively social life. A series of box-shaped benches mark the central line of the carpet; for sitting or as podiums for display of students work.

In Sweden, the City of Borås has a solid reputation when it comes to textile. Textile Fashion Center hosts different educations for designers and also a museum devoted to the subject. These activities have been relocated from other parts of the city and are since 2013 situated in an assembly of historic factory buildings, dating from the 1870s and onwards.

In the agglomerated, dense campus area separate rooms in the outdoors can be distinguished. Except for the entrance room there is a shaded space where the river forms the floor and pathways around the edges are suspended in the building facades. Yet another space is located further into the area, where a new pedestrian bridge with see-through floor connects the two banks of the river.

最近，在瑞典布罗斯纺织学校前，铺设了一个长 110m、宽 11m 的石头地毯。地毯的图案是受纺织技术的启发而设计出来的。地毯由 3 种不同的花岗岩铺设而成，它们的形状成为了这所纺织学校的象征。该处同时也是学校的主要公共空间，主要用来进行一些社交活动。此处空间还设有类似于公交站中广告牌的玻璃橱柜，里面陈列着学生们的作品。在瑞典，当纺织业出现的时候，布罗斯是享有盛名的纺织圣地。纺织时尚中心不仅为设计师们提供了一个互相交流的平台，还为他们提供了一个展示自己作品的平台。在中心设立之前，这些活动都是在城市的各个地方举办的。直到 2013 年，纺织时尚中心在一个 19 世纪 70 年代的老厂房的基础上修建完成后，所有的相关活动都转移到了此处。空间的布局十分紧凑，除了入口处有一些阴暗外，其他地方都充满了艺术色彩。河流、阶梯和道路环绕着纺织中心。为了方便参观者进出，设计师在每个进出口处都建造了步行桥。木质桌椅和木质平台为参观者提供休憩和交流的地方。

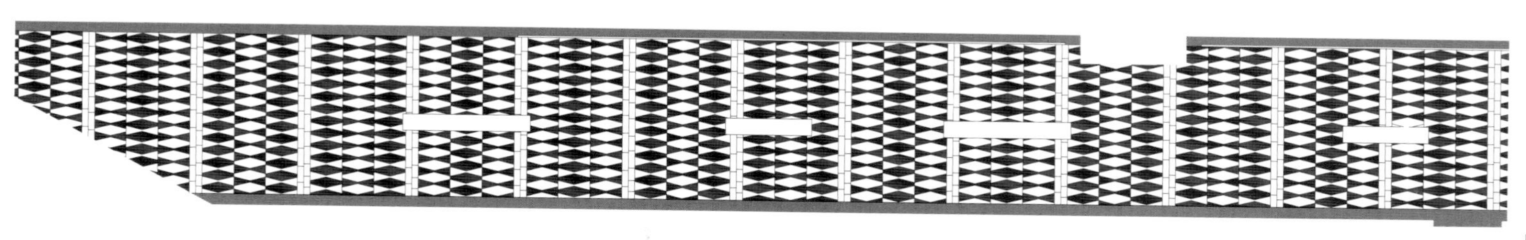

Leyteire 空间

Leyteire Courtyard

设计公司： Debarre Duplantiers Associés Architecture & Paysage, Anton Y. Olano lighting designer

地点： France（法国）

面积： 2 700 m²

摄影： Arthur Péquin, Yohan Zerdoun

The project develops four different formal elements.

1. Green world : Inserting vegetation is done through all its forms: on green walls, on flush areas, on a grid, deep in the ground.

2. Canopy: a new canopy will develop itself between the buildings, creating new views, new light filters and shadows. As a consequence, the heat effect is reduced, temperature comfort zone is improved.

3. From intimate to public show. The project is organized through a central open space with a public scene, leaving more intimate peripheral spaces on the sides. This spatial hierarchy creates a diverse thus coherent entity, able to host different functions for an urban university: shows and celebrations, group work, sunny recreational spaces.

The plan is segmented through a series of identical rectangle. This basin reflects the surrounding architecture of the 19th century, the planted trees, the breath of wind. The 80 cm wide strips integrate simple concrete furniture and planted areas, providing a welcoming shade on these central spaces. A chromatic strategy has allowed a soft and elegant atmosphere to invade this formerly delivery square to become the central spot for informal meetings in the heart of this historical university.

该项目有三种不同的要素。

1. 绿色世界：植被无处不在。墙壁上、冲洗区、网格上，甚至深入地底。

2. 顶篷：它能创造新的景观，可过滤光线，形成阴影，从而减少热效应，改善气温、提高舒适度。

3. 从私密到公开：该项目通过设置中央空地形成公共景观，为周边留出了更多私密空间。这种空间层次营造出一个多样化的统一整体，能够举办不同功能的城市活动，如表演和庆祝活动、小组活动等。

项目的空间被一系列相同的矩形而分割。周边 19 世纪的古建筑倒映在水中，树影婆娑，清风拂面。80cm 宽条状的混凝土座椅点缀在种植中，为中央空间增添了温馨的色调。这所古老学府的腹地营造出柔和淡雅的氛围，人们非常乐意在这里相聚。

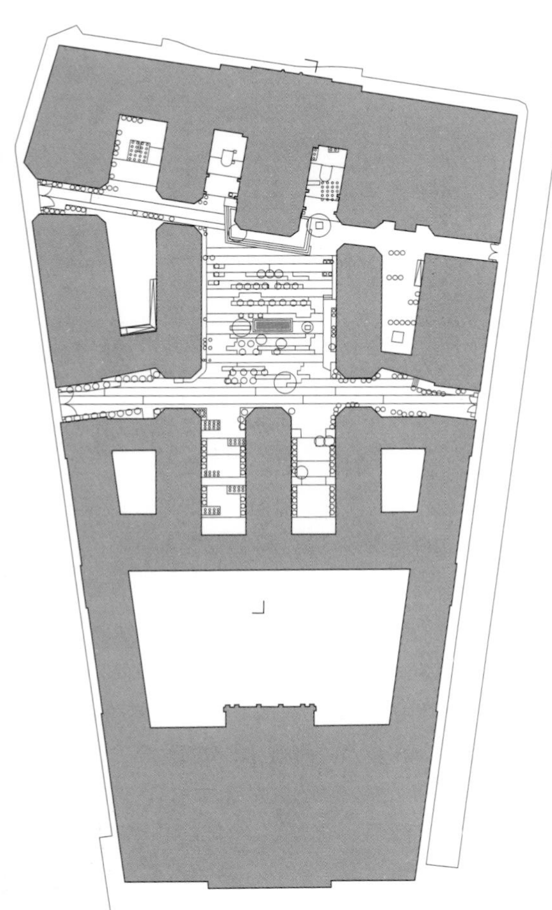

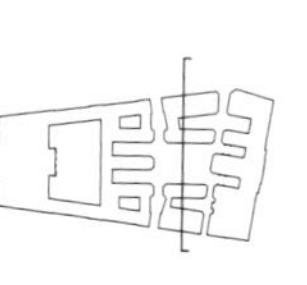

于默奥大学校园景观

Umeå Campus Park

设计公司 : Thorbjörn Andersson with Sweco architects

设计团队 : Staffan Sundström, Emma Pettersson, Mikael Johansson

地点 : Sweden（瑞典）

面积 : 23 000 m²

Umeå University is a young university, founded in the late 1960s. Here, ca 35 000 students from all over the world study in all fields of knowledge. Umeå University is located by the coast, approximately 300 km south of the Polar Circle.

A campus park should supply with a variety of designated places with the capacity to host informal discussions and exchange of ideas. It is in the open, non-hierarchical spaces, rather than in lecture auditoriums or at laboratory microscopes that the truly creative interaction between students, researchers and teachers occurs. The quality of the campus park thus enhances the attractiveness of the university as a whole.

The new Campus Park at Umeå University consists of 23 000 m² sun decks, jetties, open lawns, walking trails and terraces organized around an artificial lake. An island in the lake is the point of departure for a small archipelago with bridges leading to the southern shore. Here, the visitor meets a hilly landscape with sunny as well as shaded vales, interspersed by the white trunks of birch trees.

In the Campus Park, promenades weave themselves forward between points of social interest. These vary in size and are sometimes larger and livelier, sometimes smaller and offering intimacy. The social spots are oriented in different directions, so that the visitor always can find an attractive place. The promenades are of two types, one winding, gravelled path with shifting vistas

amigo
amigo

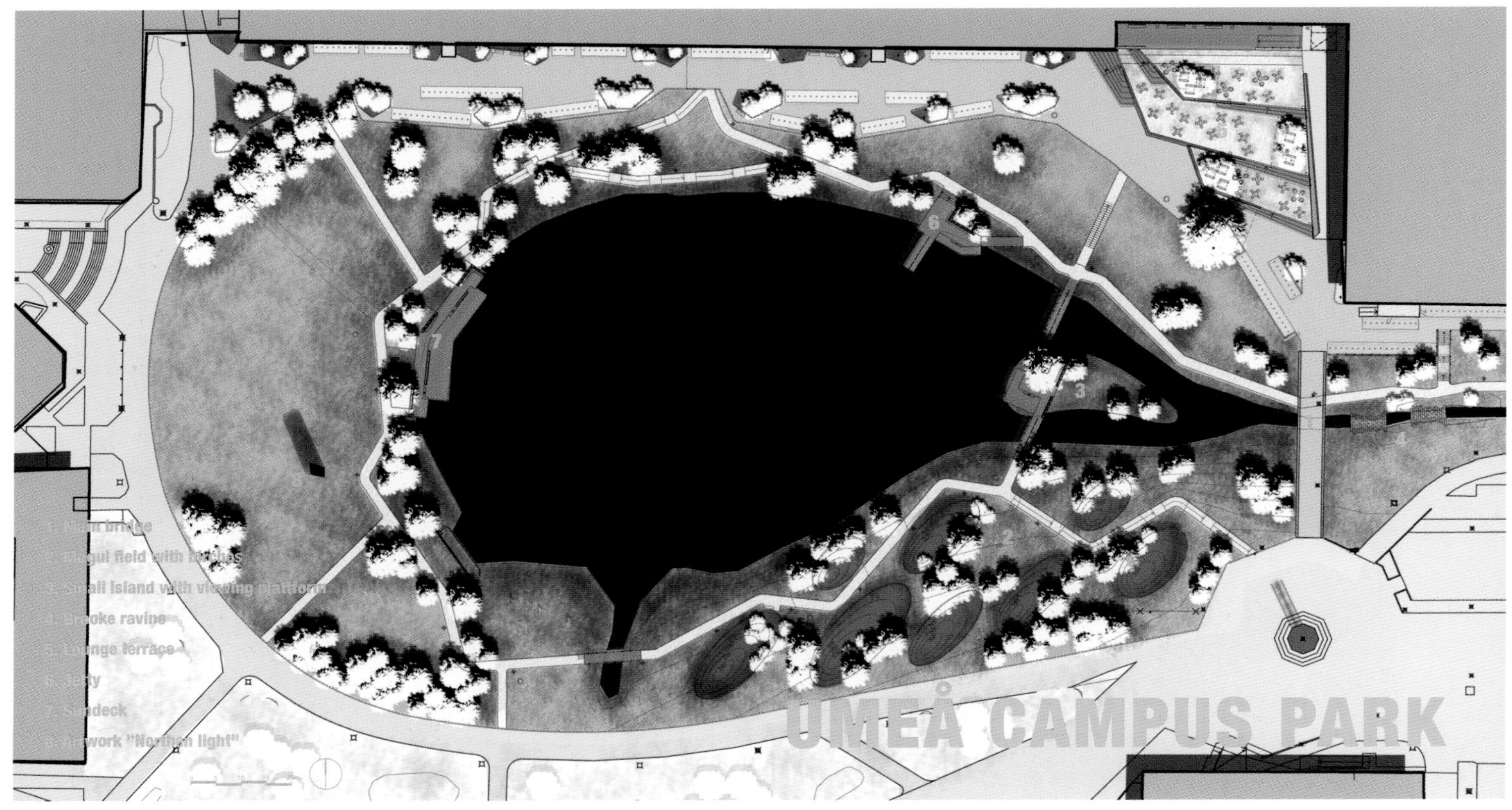

and lighting in low positions, and a wider, paved promenade which connects the entry points of the surrounding buildings.

The Corso, which is the main artery of the park, connects the main restaurant Universum with the student's union. The Corso runs on a bridge over the canyon-like affluent to the lake. Here, one finds an atmosphere almost exotic right in the Northern city of Umeå, shaded and narrow, and with a dense atmosphere along a trickling brook surrounded by large leafed vegetation.

瑞典于默奥大学建于20世纪60年代末，是一所年轻的大学。来自世界各地的35000名学生在这里学习不同的专业。于默奥大学位于海岸附近，北极圈以南约300km处。

公园校区提供不同的区域，以方便学生进行谈话和思想交流。开放、无等级的空间才能真正激发学生、研究者和教师之间的创造性互动，而并非在讲堂里或者实验室的显微镜旁。因此，公园校区的品质会从整体上增强大学的吸引力。

于默奥大学的新公园校区面积共23000m^2，阳光甲板、休闲码头、开放式草坪、步道和露台围绕着中心的人工湖。湖的一端有一座小岛，岛上有一座小桥通向南岸。在这里能欣赏到丘陵式的绿色景观、或明或暗的“山谷”和点缀其间的白桦树。

公园校区内的人行步道蜿蜒向前。它们大小不一，大的人行步道充满生气与活力，而小的步道为师生们提供了一个更为亲密的交流空间。社交空间设置在不同的方位，因此，师生们总能找到满意的地方。人行步道一共有两种，一种是盘绕式的砾石小道，可以看到远景，较低处设置了路灯；另一种是更为平坦的宽阔大道，由此可以通往周围的建筑。

该校区的主干道，连接了主餐厅和学生宿舍，以桥的形式穿过了湖泊。窄窄的阴凉步道和被浓绿植包围着的湖水为于默奥这座北方城市营造出异国的情调。

LINDELLHALLEN

以色列 Rakafot 学校

Rakafot School

设计公司： BO-Landscape Architects

设计师： Orna Ben-Ziony, Beeri Ben-Shalom

地点： Israel（以色列）

摄影： Amit Haas

The platform we designed for the school's faculty and student body to sense, observe, discover and learn about the environment facilitates the creation of a rich and changing ecosystem.

The "pond" is a depression which gathers the runoff from its immediate environs after the plentiful first rains about the time of the beginning of the school year. The pool filled up with water and a variety of flora and fauna. Besides the water acting as a decorative element in the landscape, the reservoir presents a practical lesson in collecting runoff water for irrigation and gardening as well as an opportunity to observe the wide variety of life forms which enjoy the rich habitat.

The "winter pond" is just one example of integrating the principles of green construction in designing the schoolyard of Kiryat Bialik's Rakafot Elementary School. The elementary school, with its 18 classrooms, is a pilot project of Israel's Ministry of the Environment, in advance of future construction of similar ecology-oriented schools. The extra added values accruing to this kind of school are manifold, including environmental awareness and preservation of the environment, efficient use of resources, optimum learning conditions, and environmental education.

We designed what we call the "adventure path" to provide a sense of adventure and replace the central pathway. Although it twines around the various buildings and the yard area, it is narrow, winding, and much more intriguing than the wider central pathway. It is made of asphalt, not

stone, beginning from the parking lot and passing over grassy hills and vegetation which blur the school's borderlines to create an interesting three-dimensional space. The path goes through the play spaces decorated with circles used for all types of games. In practice, we consider the pathway itself to be an extremely meaningful play environment facilitating movement and challenging the imagination.

In designing the open space, we strove to correspond with the architecture of the school buildings, with their diagonals, areas skipping over the space, and the circle patterns in the playground. The straight rows of the plantings and the elliptical platforms are complementary to the lines characterizing the school structures.

设计师为这个学校的员工和学生设计的这个能感受、观察、发现和学习的环境平台，促进了一个丰富和多样的生态系统的构成。

“冬季水池”是校园设计中遵循绿色建筑原则的一个例子。这个“水池”是一个洼地，用来收集雨水。这个水池里有各种各样的动物和植物。除了用水作为装饰元素外，这个洼地收集的雨水可以用于灌溉园艺，也为学生提供了一个观察各种生命形态的机会。

这个小学建有 18 个教室，是以色列环境部的试点项目，用来促进相似的生态学校的建造和发展。这个学校的附加价值是多方面的，包括环境意识和环境保护、资源的有效利用、最佳的学习条件和环境教育等。

设计师设计的“冒险路线”代替了中央通道，为学生营造了冒险感。它把各种建筑与庭院区域连接起来，但它是狭窄的、曲折的，比宽阔的中央通

道更加有趣。它从停车场开始，穿过草地和植被，这些元素模糊了学校的边界，建造出一个有趣的三维空间。这条通道还穿过游乐区——里面设有各种类型的游戏设施。

在设计开放空间时，设计师尽量使它和学校的建筑之间形成呼应，让建筑的对角线正对着空间，而圆形的结构则对着游乐区。沿直线排列的植物和椭圆形平台围出了学校的边界。

图书在版编目(CIP)数据

景观100+. 商业 住宅 教育 / ThinkArchit工作室主编 . —武汉：华中科技大学出版社, 2015.10
ISBN 978-7-5680-1187-7

Ⅰ. ①景… Ⅱ. ①T… Ⅲ. ①景观设计－作品集－世界－现代 Ⅳ. ①TU986

中国版本图书馆CIP数据核字(2015)第205081号

景观100+ 商业 住宅 教育 ThinkArchit工作室 主编

出版发行：华中科技大学出版社（中国・武汉）
地　　址：武汉市武昌珞喻路1037号（邮编:430074）
出 版 人：阮海洪

责任编辑：杨　睿　　责任监印：秦　英
责任校对：徐　茜　　装帧设计：张　靖

印　　刷：深圳当纳利印刷有限公司
开　　本：965 mm × 1270 mm　1/16
印　　张：19.5
字　　数：160千字
版　　次：2015年10月第1版第1次印刷
定　　价：298.00元

投稿热线：(010)64155588－8815
本书若有印装质量问题，请向出版社营销中心调换
全国免费服务热线：400－6679－118 竭诚为您服务